The A
CU00696940

The APC

Your Practical Guide to Success

JOHN WILKINSON
FRICS DIP RATING

RICS BOOKS

Material has been reproduced by kind permission of RICS.

Published by RICS Business Services Limited
a wholly owned subsidiary of
The Royal Institution of Chartered Surveyors
under the RICS Books imprint
Surveyor Court
Westwood Business Park
Coventry CV4 8JE
UK

ISBN 1 84219 121 7

1st edition published 2001

Typeset in Great Britain by Columns Design Ltd., Reading
Printed in Great Britain by Bell & Bain Ltd., Glasgow

Contents

Foreword

As Deputy Director of Education and Training (Development) of RICS I have read this book with a great deal of interest.

It contains essential guidance for all candidates enrolling onto the APC from July 2002. In July 2002 the APC was revised. Many improvements have been made to the competencies, and the second edition of this book looks at these changes in detail, while retaining much of the guidance contained in the original publication.

I believe that this is a useful guide for candidates undertaking their APC and complements the official guidance produced by RICS. It will also be a useful resource for employers, supervisors and counsellors of APC candidates.

The author, John Wilkinson has been involved with the APC for over 12 years and is a member of both the Practice Qualifications Group and the Education and Membership Committee of RICS. John has been instrumental in many of the changes that have taken place to the APC and his desire to simplify the procedures and improve the process has been a great help to candidates. He has been heavily involved in the July 2002 revisions and his detailed knowledge and experience of the APC make him the ideal person to write this practical and readable book.

I commend the advice provided in *The APC: Your Practical Guide to Success*. However, as John points out in the preface, the APC is constantly changing. Therefore, make sure you follow the most up-to-date guidance provided by RICS and regularly visit the relevant section of the website www.rics.org. This way, you will be well on your way to success.

Winifred Cooksey
Deputy Director of Education and Training (Development), RICS

Preface

In writing the second edition of this book my idea was to put my experiences of the last 12 years on paper. The book has been thoroughly revised, and I have complemented the original publication with guidance and information concerning the revisions made to the APC in July 2002. The aim of the book is to give you the correct focus and direction at the beginning of the training period, to explain the basic philosophies behind the APC (in particular, the competencies) and to guide and steer you through the training period and the final assessment interview. That's what this book is. What it is not is a quick fix to be used the night before the final assessment. There is no short cut to a properly structured training period. My belief is that it is this element which is the key to success – and so far as passing the APC is concerned, this is as close as you will come to a guarantee.

I am firmly of the opinion that the APC is first and foremost a period of training and practical experience. The July 2002 revisions reinforce this concept. If this aspect of the process is correctly put in place at the outset, and performed to the standards required, the final assessment should be a formality.

In the first edition of this book I explained that from 1 January 1997 the concept of competencies was introduced. I also commented that from that date the structure and format of the APC began to change radically. This process of evolution has continued and the various routes to professional qualification and the competencies have undergone an extensive review. In July 2002 the first editions of the all-new *candidate's guide* and *APC/ATC requirements and competencies* guide were published, together with the first edition of the *APC guide for supervisors, counsellors and employers*. In this second edition of the book, I will explain the recent changes to the APC and give you some practical guidance on how these will affect your structured training period, final assessment and the administrative and procedural aspects of the process. These matters will be covered in detail in the chapters that follow.

It is important that you understand that the APC is organic, i.e., it is growing and changing. The world does not stand still, and this fact was recognized in RICS' 'Agenda for Change', which was launched in mid-2000. The drive to raise the profile of surveyors and deliver better services globally has led to the recognition of the need for high standards of professionalism. All aspects of RICS' business activities are being thoroughly reappraised. Demands on surveyors in respect of the services provided are ever-increasing, and this in turn leads to the constant review of entry requirements into the profession, so that high standards can be maintained. It is vital that RICS invests in the correct 'seed-corn' for the future, and it is therefore necessary to keep the APC constantly under review, so that the process, format and standards can be rigorously maintained. This philosophy also applies to entry standards generally; in recent years RICS has adopted a much more dynamic approach to maintaining educational standards. I hope that you will therefore read on and read well.

I remember reading some questionnaires that had been returned by candidates after the spring 2000 assessments. One candidate commented, 'my main complaint is the endless changes to the APC procedure introduced every year'. I do hope that I have helped you to understand why this is necessary.

It is also useful for you to be aware of how changes to the APC take place. RICS is a membership organization and decisions involving change are generally made by members, sitting on the various committees that have been set up by RICS staff. So far as the APC is concerned, the committee that sits at the centre of this process is the Practice Qualifications Group. Any policy changes to the APC affecting the structure are then approved by the Education and Membership Committee, with changes to the competencies being approved by the appropriate faculty board. The Practice Qualifications Group is made up of around 10 people, representing a broad spectrum of the membership, including an international dimension. The 10 comprise experts across the various faculties, an APC doctor (usually a younger, recently qualified member of the Institution, who gives advice to candidates taking their APC), an employees' representative, an RICS training adviser (RTA), a Junior Organization representative, and others. In addition to the members who meet regularly, there are 10 corresponding members, who advise the group on a variety of areas and issues, such as the Assessment of Technical Competence (ATC – the process by which candidates acquire the technical RICS qualification), the European Society of Chartered Surveyors and Australasia. Closer to home, it would be unfair to exclude from mention my colleagues and some close friends in Northern Ireland and Scotland, who also provide advice.

I would like to highlight the excellent work that is being carried out at the moment by the Practice Qualifications Department of RICS (the team of staff who support and advise the Practice Qualifications Group). As RICS expands globally, so must the APC. The APC is indeed 'going global' and is being set up in many countries in Europe, including Belgium, France, Germany, the Netherlands, Greece, Hungary and Cyprus. It is also being established further afield, in Australia, Singapore and China. This globalization obviously adds to the drive for change, and while I have sympathy with the candidate who was frustrated by 'the endless changes', you need to think of these developments in terms of the prospects and job opportunities that are opening up for you around the world!

One final comment – all current matters and developments regarding the APC are published on the RICS website (www.rics.org). Hard copies of this information can also be obtained from RICS (contact the Practice Qualifications Department on 020 7222 7000).

An overview of the APC

This chapter provides an overview of the basic philosophies and key concepts of the APC, as well as an introduction to the various guides and forms that need to be completed, and a note of the essential dates and deadlines that occur at key stages within the training period. It outlines the roles and responsibilities of some of the people essential to your training and development, such as RICS training advisers (RTAs), APC doctors, your employer and RICS itself.

Before you begin the APC, you must choose an APC 'route'. On completion of the qualification, this enables you to join the RICS faculty of your choice. There are two types of route: the expert (or specialist) route (applying for example, to the Antiques and Fine Arts faculty) and the mainstream routes (applying to Building Surveying, for instance). This book aims to guide you along all of these routes, to success in the final assessment.

OFFICIAL GUIDANCE

This book will not replicate the two official RICS guides to the APC: the *candidate's guide* and the *APC/ATC requirements and competencies* guide. I suggest that you read this book first and then the two guides. It is essential that you do read and study these guides. I still come across too many candidates who have not read them, particularly the *candidate's guide*. This fact is made quite obvious to me by some of the questions I am asked, such as, 'why do I need to know anything about the Rules of Conduct, when I have no direct experience or involvement?', or 'how long after the final assessment will I receive my results?' or 'if I am referred, what happens to my critical analysis?'. All of these issues are covered in the guide, so beware!

You should also be aware that this book focuses on the first editions (July 2002) of the *candidate's guide* and the *APC/ATC requirements and competencies* guide. These guides will generally apply to candidates enrolling onto the APC after that date. Candidates already in the system should follow the details of the guides that were in existence when they enrolled. For example, the last of the previous set of guides (edition five, published in January 2001) will be applicable to those candidates who enrolled between January 2001 and June 2002. However, I do need to issue a warning. You must keep up to date with any changes that are published by RICS and, most importantly, watch out for any transitional arrangements that may occur. It is likely that at some point in the future all candidates will be moved onto the July 2002 routes of qualification and competency requirements.

Another change that took place in July 2002 was the introduction of the *APC guide for supervisors, counsellors and employers*. This is aimed at providing employers with all the help and guidance needed to support a candidate through the APC. Above all, it outlines the commitment needed – and in this respect it is important that your employer not only has a copy of the guide, but is fully aware of his or her responsibilities. A little encouragement from you will not go amiss.

WHAT IS THE APC?

Let's start with the basic philosophies: what is the APC? Quite simply, **the APC is the process by which RICS seeks to be satisfied that candidates who wish to become members of the Institution are competent to practise as chartered surveyors.** To demonstrate this competence, candidates will undergo a rigorous and demanding period of structured training, over a minimum period of 24 calendar months, in which they must gain a minimum of 400 days of relevant experience. The objective is to enable the knowledge and theory gained primarily in further education to be complemented by practical experience. The second part of the APC process is then the final assessment interview.

The APC therefore comprises two components:

1. a period of structured training: during this period you should keep a record, in a diary, of experience gained. You must also keep a log book and a record of progress. The log book is a summary of the experience contained in the diary grouped together under the various competencies. The record of progress is the document in which you record your progress of achievement against the various competencies of your chosen route. The training period also incorporates your professional development, an outline of which is given later in this chapter, and in more detail in chapter 2;
2. the final assessment interview: a panel of three practitioners (assessors) will interview you over a period of one hour, and form a judgement, or 'assess', whether you are competent to practise as a member of RICS.

The final assessment interview is dealt with later in the book. The intention here is to look at the key concepts that comprise the training period. However, before doing this, I would like to expand your understanding of what the APC is all about. Think of the APC as the practical training and experience which, when added to the full-time study carried out at university, leads to membership of RICS. This concept is illustrated in figure 1.

The illustration in figure 1 is not exactly complex, but I hope it conveys my meaning. The first part of the equation can include many variants, which will also involve practical experience – for example, a degree course at university which contains a sandwich year.

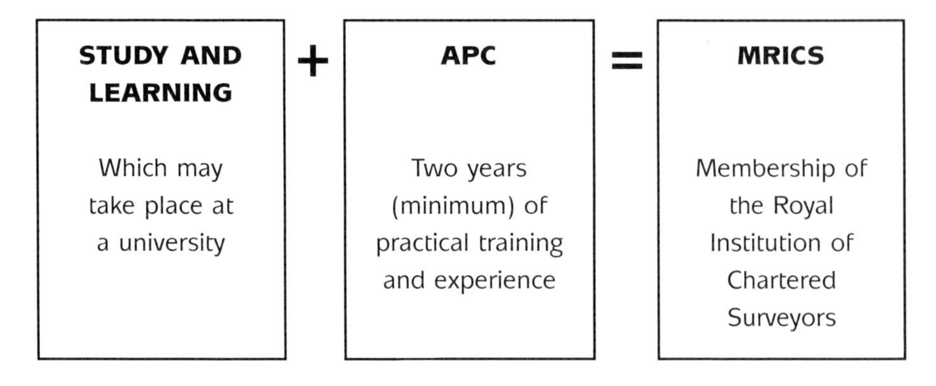

Figure 1 Route to membership of the RICS

A commonly held belief is that the 'assessment' is all about the questions asked in the final assessment interview. This is not true. The interview panel do not simply base their assessment on the questions and answers in the interview, but take into account all other aspects of the two-year period that precedes the interview: the depth and breadth of training; the quality of the documents and written reports (including the critical analysis and professional development record); and your presentation at the interview. All these matters comprise the 'assessment'. At the end of the interview the panel will view your training and development 'in the round', by taking a holistic view of your performance.

KEY CONCEPTS AND DOCUMENTS

Competent to practise

The APC is first and foremost a period of structured training and practical experience, which culminates in the final assessment interview, the objective of which is to assess your competence to carry out the work of a qualified surveyor. To be 'competent' is to have the skill or ability to perform a task or function. This ability can vary from being merely able, to being expert in a particular sphere of activity. When considering ability or expertise, do not think about this concept as simply relating to the surveying profession. Changing a flat tyre on a car or preparing an evening meal also demands a level of ability or expertise.

Competent to practise: streamlining the final assessment

In the context of the surveying profession and the APC, the Practice Qualifications Group instigated some research in 1999, with a view to further defining and clarifying the meaning of 'professional competence', for the purpose of streamlining the decision-making process in the final assessment. This research resulted in the drawing together of the many mandatory, core and optional competencies under five broad headings:

1 knowledge;
2 problem-solving ability;
3 Rules of Conduct and professional ethics;
4 business- or practice-related knowledge and skills;
5 personal and inter-personal skills.

The idea behind this streamlined approach is to enhance the consistency of the final assessment decision by providing the assessment panel with a clear picture of what 'competent to practise' looks like. In brief, at the final assessment the panel will be looking for you to demonstrate that you meet with the knowledge requirements of your chosen route, and that you can use this knowledge to solve practical problems. You will also have to show that you have developed business skills, an appropriate range of personal and inter-personal skills, and most importantly, that you are aware of and intend to act in accordance with the Rules of Conduct and ethical standards laid down by RICS. It is worth noting that the July 2002 *APC/ATC requirements and competencies* guide devotes a section to consideration of the Rules of Conduct and professional ethics (page 5, in Ethics, professional identity and accountability).

A competency

A competency is a statement of the skills or abilities required to perform a specific task or function. It is based upon attitudes and behaviours, as well as skills and knowledge. For the APC, the requirements and level of attainment for each route are set out in the *APC/ATC requirements and competencies* guide. Chapter 2 will develop your understanding of the specific competencies of your chosen route and show how, at the final assessment, these will be drawn together by the assessment panel using the streamlined approach referred to above.

Faculty statements

In the first edition of the *APC/ATC requirements and competencies*, the competencies are written in a generic format, without any technical detail. The detail is contained in the 'faculty statement' that immediately precedes the detailed requirements of each route. It is vital that you read these statements. They are not only important with regard (in many routes) to your initial choice of optional competencies (See chapter 2), but will also provide you with an idea of the breadth and depth of technical knowledge that you need to acquire during the training period, in readiness for the questions that will be asked at your final assessment. Importantly, your initial choice of optional competencies will itself be judged by the panel at your final assessment.

Structured training

'Structured training' is, as its name suggests, a structured approach to the delivery of training over a given period. From 1 October 2000 it has been mandatory for all firms registering new APC candidates to have a structured training agreement in place. In addition, candidates must attach a 'competency achievement planner' to their application form, showing the training that has been planned for them.

A template and examples of a structured training agreement and competency achievement planner can be found at the back of the *candidates guide*.

The structured training agreement and competency achievement planner are simply documents that formalize the intention of the parties to deliver (on the part of the employer) and receive (on the part of the candidate) the training requirements of the chosen route over an agreed period and to specified levels of competence. You will find more details on these documents in chapter 2.

Note that there is no minimum requirement of training days to be completed under each competency, other than the overall requirement of the structured training period. This is usually a minimum of 400 days within 24 calendar months.

Diary

The diary is a day-to-day record of how you have been building your experience. The detail contained in it will assist you in completing your log book and record of progress.

Log book

The log book forms a monthly summary of the entries in your diary and provides a total of the number of days of experience gained in each of the competencies.

Record of progress

The record of progress charts your progress against the competency requirements of your chosen route. In other words, it is a record of attainment, which is certified by your supervisor and counsellor. There is a series of forms to be completed: the three-monthly supervisor's reports; the six-monthly counsellor's reports; the interim and final assessment records; and the referred candidates' form. These can be found in the templates at the back of the *candidate's guide*.

Professional development

Another important aspect of the training period is the requirement for you to undertake a minimum of 48 hours of professional development per annum. This is to provide you with the opportunity to gain knowledge and skills that might not be available in your day-to-day training environment. It may be used to complement the requirements of the mandatory competencies. Professional development is recorded in your record of progress, with a note of the nature of the training, the date of the event and the number of hours carried out.

In the context of life-long learning, your professional development never ends. When you have completed your professional development for APC purposes, RICS then requires you, as a member, to undertake a minimum number of hours of continuing professional development (CPD) thereafter. The concept of professional development is explored in more detail in chapter 2.

Change of employer

If you change employer during the training period, your records will be continued in the usual way, but there must be a clear note of the change of employer. In particular, the new position regarding your supervisor and/or counsellor must be indicated for the purposes of certification at the final assessment. RICS must also be advised of the change of employer. (To do this you must complete and return the change of employer form to RICS. This form will be sent to you when you enrol on the APC.) Upon receiving this form, RICS will send you a letter either approving the move or detailing the action you need to take to resolve any problems.

Having considered the basic philosophies and some of the documents of the APC, I now want to take a brief look at some of the important dates and deadlines to which you will need to adhere during the training period.

IMPORTANT DATES AND DEADLINES – A CHRONOLOGY OF THE APC

Enrolment

Enrolment can take place at any time, by approaching RICS for an application pack. However, it is important to note that you cannot backdate the recording of experience. You can only begin recording experience from the date your completed application form is received by RICS.

Once you have approached RICS, you will be sent an application pack, which will contain the *candidates guide*, the *APC/ATC requirements and competencies* guide, the *APC guide for supervisors, counsellors and employers* and an application form. This agreement need not be sent to RICS, but should be kept by your employer and made available on request to your RTA. However, your competency achievement planner, which summarizes the training proposed by your employer, must accompany your application form. There is one exception to this rule. If your employer has a structured training agreement that has been discussed and agreed with an RTA – that is, an 'approved' structured training agreement – you do not need to send your competency achievement planner to RICS with your application form. However, the planner must still be completed, used and reviewed throughout your training period.

Acknowledgement by RICS

Acknowledgement by RICS will usually occur within two weeks of your application being received. You will be given a date from which you may start recording your experience. Don't forget the importance of this date: before you reach the final assessment, you will need to have completed a minimum of 400 days' experience within 24 calendar months. To be considered for either a spring or autumn assessment, you must have completed the required minimum training period. It follows, therefore, that a delay of a few weeks in enrolling could put the assessment date back by six months. You will also be given a date for your expected final assessment, a form to be completed and returned should you change employer, and a copy of the Rules of Conduct. You should take care to familiarize yourself with these Rules, as your knowledge of them will form part of your final assessment interview (see chapter 5 for more details).

Three-monthly reviews by your supervisor

Every three months your supervisor will discuss and review your progress against the competencies and complete a progress report.

Six-monthly reviews by your counsellor

At six-monthly intervals your counsellor will discuss and review your progress against the competencies and complete a progress report. This report constitutes a second opinion to that of the supervisor and will be completed in conjunction with you and your supervisor.

Interim assessment

The interim assessment must be completed within one month of the first 12 months of training. The format for this is outlined in the *candidates guide*. It is important to note that a minimum of a further

12 months of training must be completed before you can sit the final assessment. It is also worth mentioning that the interim assessment may be the subject of audit during a visit by an RTA.

Final assessment

The final assessment application pack will be sent to you by RICS approximately five months before the final assessment dates (assessments are held twice a year). You will need to send in the completed application form within the dates specified in the application pack. You will then have approximately one month to submit the required documents for the final assessment presentation and interview. A list of these documents can be found in the *candidates guide* under the heading 'Pre-Assessment Submissions'.

Results

Results will be posted within 21 days of the interview date.

Appeals

An appeal, if necessary, should be received by RICS no later than 14 days from the date your result was posted.

FINDING HELP

There may be occasions during your training and in the run-up to the final assessment when you need help, in the form of advice or guidance, over and above that given by your employer. There are a variety of sources that you may approach:

- APC helpline: RICS provides a helpline on all aspects of the APC, from start to finish and beyond (Tel: 020 7334 3886 or e-mail apc@rics.org.uk);

- APC doctors: the APC doctor scheme is available locally. APC doctors are usually younger, recently qualified members, who can advise and guide you through the APC, based on their own experience of the qualification. Details of APC doctors in your area can be obtained from RICS;
- RTAs: early in 1997, with the advent of the competency-based APC, RICS set up a national network of regionally based training advisers. The group comprises a body of members from the faculties, all of whom have surveying, teaching or training experience. The RTAs are employed by RICS and, while their principal function is to advise employers on structured training schemes, they are available to advise candidates on all aspects of the APC;
- RICS website: the RICS website (www.rics.org) contains details on the APC. This is regularly updated and provides a great deal of advice and guidance. It is particularly useful for keeping up to date with changes to the APC, but also contains a host of other useful pieces of advice and information that may be of interest.

CHOOSING A PROSPECTIVE EMPLOYER

It would not be appropriate for me to close this chapter without reference to the employer. I started the chapter by considering the basic philosophies behind the APC, by examining the concept of being 'competent to practise', and exploring the meaning of this in the context of structured training and your progress against the competencies. I believe that this is only one side of the coin. Fundamental to the whole process is the role of your employer – and my advice to all candidates in this respect is to look before you leap. There is a saying in the North East that 'shy bairns get nowt' – so do not be frightened to ask a prospective employer what it is they have to offer you by way of a structured training scheme. Questions you should ask include the following:

- does the employer have a structured training scheme approved by RICS?

- is there a proven track record of training APC candidates?
- does the employer have its own schemes for providing professional development?
- does the employer have Investor in People accreditation? The Investor in People standard is given to firms who set high standards in terms of their commitment to, and their planning, delivery and evaluation of training and development for their staff;
- on a more day-to-day basis, how does the employer plan to assist you in meeting the competency requirements? Can everything be done in-house? Will you be moved between departments or offices? Are secondments or exchanges available with other firms to make up any shortfalls? In addition, you should ask if you can speak to any of the firm's current or recent APC candidates;
- who will be your supervisor and counsellor, and what is their experience of the APC?

Patience and understanding is also recommended. Most candidates have an expectation of being ready for the final assessment as soon as the minimum training period of 400 days within 24 calendar months has elapsed. This is not always possible. Sometimes, because of business practicalities or priorities, your employer may not have been able to give you the full spread of training against the competencies. Alternatively, you may not have progressed against certain competencies as quickly as expected. Don't forget, we all develop at different speeds in any of life's skills or functions. My advice is to apply for your final assessment only when you are ready and feel confident that you have gained the necessary training and experience, even if that means that your training period lasts for 30, or even 36, months. Be prepared to co-operate with your employer and, most importantly, listen to the advice of your supervisor and counsellor.

Remember that your employer does not exist merely to provide APC training for you. There is a business to run and profits to be made. This issue will prove a key challenge for most candidates. While you will develop certain expectations concerning your APC training, your employer will also be looking for a return from you by way of hard work, enterprise and fee income! You will need to approach the partnership with your employer with a degree of understanding and awareness, giving careful thought as to how you will balance the demands of your employer with your own need to make progress with the APC.

Finally, the importance of the employer cannot be over-emphasized in terms of the appointment of your supervisor and counsellor, who, generally speaking, will be part of your employer's team. Your supervisor and counsellor will act as guides and mentors throughout and in a lot of instances beyond your training period and their roles will be of particular importance during preparation for your final assessment. It is important that this is an active partnership, where the employer takes a proactive approach, by way of interest and involvement, in your training and development.

For some time I have listened to the debate that the large firms provide the best opportunities for APC training and that small firms may struggle. I cannot agree with this statement. In my experience, a small firm with a varied portfolio of work can provide more than adequate training and experience for APC candidates. I have seen many examples of this over the years.

So do not be slow in coming forward, in your own best interest, to find an employer who will provide you with the training and experience you need, to ensure that success at the final assessment interview is a formality. But do not lose sight of the fact that you are entering into a partnership and that your employer will also want a return on the investment.

SUMMARY

- It is essential to read the *candidate's guide* and the *APC/ATC requirements and competencies* guide.
- You will need to complete a minimum two-year period (400 days within a minimum of 24 calendar months) of training and practical experience.
- It is important that you have a clear picture of what is meant by the term 'competent to practise', with regard to knowledge; problem-solving abilities; business and practice skills; personal and inter-personal skills; and adherence to the Rules of Conduct and professional ethics.
- A structured training agreement must be completed by your employer; you must also complete a competency achievement planner.
- Training and practical experience culminates in the final assessment: a one-hour interview with a panel of three experienced practitioners, who will consider whether you are 'competent to practise'.
- Information and evidence of training and experience must be kept in your diary, log book and record of progress.
- You must undertake a minimum of 48 hours of professional development for every 12 months of training completed.
- If you change employer, you must notify RICS.
- Experience only counts from the date on which your completed application form is received by RICS. RICS will advise you of this date.
- Your application to enrol must be accompanied by a competency achievement planner (unless your employer has an RICS-approved structured training agreement).
- The interim assessment must be completed within one month of the first 12 months of training; a further 12 months of training must be completed before you are eligible for the final assessment.
- Approximately five months before the final assessment dates, RICS will send you an application pack.
- Results of the final assessment interview will be posted within 21 days of the interview date. If you are referred and wish to appeal, you must do so no later than 14 days from the date your result was posted by RICS.
- Help with the APC is available from RICS, APC doctors and RTAs.
- When choosing a prospective employer, 'look before you leap' and remember the partnership aspect.

The training period

This chapter looks at how to go about managing your training period. It aims to develop your understanding of the competencies and illustrate how professional development can be used as a tool to assist your progress. Structured training will be considered in some detail, with particular regard to the various forms that need to be completed.

THE COMPETENCIES

A key aspect to the successful management of the training period is an in-depth knowledge and understanding of the competencies and requirements of each RICS faculty or surveying specialism, which I refer to as 'APC routes'. It is therefore appropriate that this issue is the first to be considered in this section of the book, as it will get you started on the right track and lead towards success at the final assessment.

In chapter 1 we considered the meaning of the word 'competent' and explored what is meant by a competency. In the broader context, the use of competency-based training and, most importantly, competency-based interviewing, is fast becoming a worldwide phenomenon. As your career progresses and develops, you will find that your appreciation of this subject will grow. However, at this stage it is important that you fully understand competencies in the context of your APC. The guidance that follows is therefore tailored to fit that particular need.

Your point of reference is the *APC/ATC requirements and competencies* guide, which sets out what you need to achieve by way of skills and abilities over the training period (the guide also contains the competency requirements for the technical candidates). This document can broadly be divided into two parts. The first part indicates the specific competency requirements of each route, with guidelines on the number of competencies to be covered during the training period, and the depth and level of attainment required. The choice of optional competencies, from the 'full list', is at the discretion of the candidate.

However, your choice must reflect not only the work you carry out in your day-to-day environment, but your chosen faculty as well. Most importantly, this choice will be considered as a reflection of your judgement at the final assessment.

It is therefore vital that you follow the guidance contained in the faculty statements that precede the requirements of each route. The second part of the document sets out the 'full list' of competencies in alphabetical order, with a reference for each – for example, Cadastre and Land Management Ref 009. This reference number is for use on your record of progress. It should be noted that the Rules of Conduct competency is not in the alphabetical list and therefore does not have a reference number. A heading on the Record of Progress will suffice to identify this competency.

Candidates opting for the Management Consultancy route should note that the optional competencies are listed in the requirements of the route, as opposed to being taken from the full list of competencies. However, it is possible to make substitutions from the full list. These special rules should be considered carefully.

The competencies are varied and cover a wide range of technical, professional, business, personal and inter-personal skills. For each competency, there are three levels of attainment:

Level 1 – knowledge and understanding of the areas covered by the competency;
Level 2 – application of this knowledge and understanding in normal practical situations;
Level 3 – ability to provide reasoned advice, or to apply knowledge and understanding, in more complex technical situations.

Consider the example below of the building surveying competency of ‘building pathology’, which progresses in complexity across the levels.

Building pathology

Level 1: demonstrate your knowledge and understanding of building defects;
Level 2: apply your knowledge to undertake surveys, and to diagnose causes and mechanisms of failure;
Level 3: give reasoned advice, and prepare and present reports.

A typical final assessment question for this competency might be something along the following lines:

> ‘I notice from your summaries of experience that you have carried out building surveys on a number of traditionally built 1960s houses. Describe how you went about those surveys [level 1]. What were the common types of failure to brickwork observed while carrying out those inspections? What were the causes and how did you diagnose these? [level 2]. What method of repair did you recommend to your client, and why? What sort of issues did you include in your final report and recommendations? [level 3].’

You will reach these levels in a logical progression and in successive stages. There is no minimum number of days of training required to achieve a particular level of attainment in any of the competencies. In discussion with your supervisor and counsellor, a decision will be made as to when you have reached the required level of skill and ability in any particular competency. The number of days taken to reach the appropriate level will be dependent upon a combination of the following factors:

- your starting point – have you had any previous experience?
- your aptitude and speed of learning in the competency;
- the quality of the training and experience given by the employer;
- the particular competency – is it measurement, or nuclear physics?!

A judgement as to your achievement will be made by your supervisor and counsellor, who will then sign you off in the appropriate column in your record of progress.

Each route requires you to satisfy three types of competency:

1 mandatory competencies;
2 core competencies;
3 optional competencies.

The mandatory competencies generally relate to personal, interpersonal and business skills and are common to all routes. The core competencies relate to the primary skills of your chosen route. The optional competencies are selected by you.

The mandatory competencies

The minimum standards for mandatory competencies set out in the *APC/ATC requirements and competencies* guide have been set by the Practice Qualifications Group. These competencies are structured in levels. You must achieve the minimum standards, as follows:

- ethics, professional identity and accountability (to level 3);
- conflict avoidance, management and dispute resolution procedures (to level 1);
- collection, retrieval and analysis of information and data (to level 1);
- customer care (to level 2);
- environmental awareness (to level 1);
- health and safety (to level 1);
- information technology (to level 1);
- law (to level 1);
- oral communication (to level 2);
- self-management (to level 3);
- teamworking (to level 1);
- written/graphic communication (to level 2).

In addition to the above, all APC candidates must achieve two further business management and inter-personal skill competencies, to level 1, from the list below:

- accounting principles and procedures;
- business management;
- corporate and public communications;
- leadership;
- managing people;
- managing resources;
- negotiating skills;
- recruitment and selection.

All of the above competencies (with the exception of the ethics, professional identity and accountability competency) are defined in an alphabetical list in *APC/ATC requirements and competencies*.

If you are an APC candidate following an expert route to membership, you must achieve two further business management and interpersonal skill competencies, to level 1, from the list below:

- accounting principles and procedures;
- business management;
- corporate and public communications;
- negotiating skills;
- recruitment and selection.

In addition, candidates following an expert route must achieve the following three competencies:

- leadership (to level 2);
- managing people (to level 2);
- managing resources (to level 2).

The minimum standards described above may also be included at a higher level, if appropriate to the route of the faculty or surveying specialism. If a faculty has included a mandatory competency to a higher level, this will also appear in the core competency list for that route, to the higher level.

A level 1 mandatory competency may also be chosen as an optional competency at level 3, in routes where it is not also a core competency. It is important that the advice contained in the *APC/ATC requirements and competencies* guide is taken literally. 'A level 1' means 'one level 1 mandatory competency' – that is, you may only substitute one mandatory competency to a higher level as an optional competency. My advice is that wherever possible, you stick to the technical skill competencies that relate to your chosen route.

The reason for this, as explained above, is that your selection of optional competencies will be considered as a reflection of your judgement at the final assessment. I would suggest that if you use a mandatory competency as an optional competency at a higher level, the competency must be of high importance in your day-to-day work.

I will deal with structured training later in this chapter. Overall, my advice regarding mandatory competencies is that when considering your structured training agreement, you should sort out how you propose to achieve the required core and optional competencies, and then sift through the list of mandatory competencies. Identify the mandatory competencies where you will receive the appropriate training within your day-to-day work, as part of your structured training. This will leave you with a remainder which will require you to consider additional learning or a specific training course. These actions will qualify for your professional development (covered in the next section of this chapter).

The above only deals with the competencies with reference to the training period. I will return to the competencies when we look at the final assessment interview later in the book. At this stage, I merely want to note that the final assessment interview is a part competency-based interview, for which there are specific skills required of both the assessment panel and the interviewee. In a full competency-based interview the panel would not usually question you outside of your main areas of training and experience. However, in a part competency-based interview, some questions may relate to matters of a more general nature. This issue is covered in more detail in chapter 5.

PROFESSIONAL DEVELOPMENT

For each 12 months of practical training that is completed, you must undertake an annual minimum of 48 hours of professional development. The idea behind professional development is that it gives you the opportunity to acquire some of the additional skills and knowledge that it will not always be possible for your employer to provide within the week-to-week business of the practice. This may particularly apply to the various mandatory competencies referred to in the previous section of this chapter.

An important aspect of your professional development is that it should be planned and structured in such a way as to remain flexible. It should be designed to complement and support your training and development in the context of the various competencies. Professional development may comprise formal training courses or more informal types of learning, such as structured reading, distance-learning or e-learning programmes, and secondments. It is important that you accept ownership of your professional development, recognizing that the planning, acquiring and evaluating of it is your responsibility. Note that the *candidate's guide* also provides excellent guidance on professional development.

A typical annual professional development plan could look like the one presented in figure 2.

Professional Development for 2003

Technical skills development: linked to core/optional competencies – normally 16 hours.

Skills development: linked to mandatory competencies – normally 16 hours.

Professional practice skills development: linked to those competencies associated with professional practice, ethics and conflicts of interest – a further 16 hours.

Figure 2 Typical annual professional development planner

On a practical note, make sure that your professional development complements your structured training agreement, and ensure that at the interim and final assessments you can provide evidence of a planned and systematic approach. There should be a clearly defined relationship between the topics selected and the competencies. If you feel that there is a need for variation regarding the number of hours allocated, discuss this with your supervisor and counsellor and include an explanation of this departure from the norm in:

- your interim summary of training and experience, as part of your interim assessment (see chapter 3);
- your final summary of training and experience, as part of the final assessment submissions (see chapter 4).

STRUCTURED TRAINING

In chapter 1 we looked briefly at the concept of structured training. In the context of the APC, this is training that is discussed, planned, reviewed and, if necessary, revised, and which, most importantly, forms the basis of an agreement between the parties. This section of the book explores this in more detail.

So far as the APC is concerned, there are two important components of your structured training period:

1. the structured training agreement – setting out the detail of your employer's company policy;
2. the competency achievement planner – setting out your plan for the training period, in terms of the competencies that you are seeking to achieve, when the appropriate levels of attainment will be reached and how you propose to reach the required level.

The development of a structured training agreement is mandatory for all candidates and is prepared by your employer. It does not have to be sent to RICS, but should be kept by your employer and made available on request to your RICS training adviser (RTA). The competency achievement planner, which, in effect, summarizes the structured training agreement, must accompany your application form for enrolment on the APC. If it does not, you will receive a reply from RICS putting your enrolment on hold and giving you 20 working days in which to send the planner. There is, however, one exception to the rule. If your employer has held discussions with an RTA over the company's structured training agreement, and if this has been approved by the RTA, you do not need to submit a competency achievement planner with your application to enrol.

Structured training agreement

Guidance concerning the preparation of a structured training agreement can be found in the various templates at the back of the *candidate's guide*. You will probably find that most employers who have recently trained APC candidates will already have a training agreement in place and will be familiar with RICS requirements. All that will be needed is a degree of tailoring to suit you and your proposed training, in relation to time-scales and competencies. However, if this is not the case, the guidance in the *candidate's guide* will provide a useful discussion document for you and your employer, when planning and agreeing your training for the next two years.

The *candidate's guide* provides guidance only; your employer may wish to tailor an agreement more suited to his or her firm. The *candidate's guide* suggests that the following information should be included:

Competency Achievement Planner

Name: Samantha Jackson

Company name: Smith and Co Chartered Surveyors

Ref. no		Qtr 1	Qtr 2	Qtr 3	Qtr 4	Qtr 5	Qtr 6	Qtr 7	Qtr 8
Mandatory competencies									
Ethics statement in *APC /ATC guide*	Code of conduct, professional practice & bye-laws								
A015	Conflict avoidance, management & dispute resolution procedures								
A012	Collection, retrieval & analysis of information & data								
A026	Customer care								
	etc. See full list of requirements and levels in *APC/ATC requirements and competencies* and note any additional requirements of your route.								

Figure 3 Example of a competency achievement planner

Competency Achievement Planner

Name: Samantha Jackson

Company name: Smith and Co Chartered Surveyors

Ref. no		Qtr 1	Qtr 2	Qtr 3	Qtr 4	Qtr 5	Qtr 6	Qtr 7	Qtr 8
Core competencies									
00?	Competency 1	**Level 1**		**Level 2**			**Level 3**		
00?	Competency 2	**Level 1**		**Level 2**			**Level 3**		
00?	Competency 3	**Level 1**		**Level 2**		**Level 3**			
00?	Competency 4			**Level 1**		**Level 2**		**Level 3**	
Optional competencies									
00?	Competency A			**Level 1**		**Level 2**			
00?	Competency B				**Level 1**		**Level 2**		
00?	Competency C				**Level 1**		**Level 2**		

Commentary

Figure 3 (Cont'd) Example of a competency achievement planner

- guidance which refers to the minimum time periods and to the mandatory records (your diary, log book, record of progress and supervisor's and counsellor's progress reports);
- a candidate's statement, outlining your commitment, together with a record of the relevant dates from registration to the final assessment;
- an employer's statement, outlining his or her commitment and providing details of the firm in terms of activities, work areas, staffing levels, the geographical location of offices, etc;
- a note of the employer's policies concerning payment of fees, leave arrangements for APC, professional development and referred candidates;
- details of the supervisor and counsellor;
- a commitment to all experience being recorded in line with the guidance set out in *APC/ATC requirements and competencies;*
- a diary of relevant dates, particularly noting the three-monthly reviews with the supervisor and six-monthly reviews with the counsellor.

Competency achievement planner

As mentioned above, the competency achievement planner sets out your plan for the training period. Detailed information on competency achievement planners can be found in the guidance in the RICS application pack. An example of a competency achievement planner is shown in figure 3.

In keeping with the philosophies outlined earlier in this chapter regarding the mandatory competencies and professional development, your development against some of the competencies – for example, business skills – will not have a start or end date. Development will continue throughout the training period and will form the start of your commitment to life-long learning, or CPD.

Your competency achievement planner should be accompanied by competency monitoring tables – one for each competency. An example of such a table can be found in the templates in the *candidate's guide*. Each table will allow you to focus on a single competency and will provide space enabling you to review training completed and to forward plan the training and development that is still required to complete the competency to the appropriate level.

This chapter has explained in detail the principle of competencies, including professional development, in the context of structured training, and also the various issues that need to be considered at the beginning of the journey towards the final assessment. The objective is to have an exciting and enjoyable period of travel, in which you will learn a great deal and set the foundations upon which to build a successful and rewarding career. In the final assessment, all of the detail and information built up in the training period will be considered by the assessment panel and used to target questions during the interview. It is therefore also important that you are aware of how this information will be used at the end of the journey. This will be covered in detail in chapters 4 and 5. However, it is worth pointing out here that in the context of the final assessment it is imperative that you give some thought as to how to manage and record information during the training period.

SUMMARY

- Competencies are written to three levels and are generally progressive in terms of skills and abilities.
- There is no minimum requirement for the number of days needed to achieve each competency. The level of attainment is decided by your supervisor and counsellor in discussion with yourself.
- Mandatory competencies are compulsory for all candidates.
- The final assessment interview is part competency-based. Some questions may relate to general matters outside of your main areas of training.

- Professional development should be linked to and used to complement your structured training agreement. At the final assessment your documentation must show evidence of a planned and systematic approach to professional development.
- Structured training is training that is discussed, planned and reviewed between the parties to the agreement. There are two main documents:
  1) the structured training agreement; and
  2) the competency achievement planner.

  The competency achievement planner should be accompanied by a series of monitoring tables – one for each competency.
- Don't forget your supervisor and counsellor declarations on your training agreement and competency achievement planner.

Information management

This chapter looks at the various records that you need to keep during the training period. It provides advice and guidance on how to keep these records and considers their importance in the context of the final assessment. The role of the supervisor and counsellor, particularly regarding the interim assessment, is also covered.

WHY KEEP RECORDS?

You should be aware of the importance of the various records you need to keep in the context of the final assessment. The information contained in your log book, professional development record, record of progress and various reports will not only provide the assessment panel with the evidence that you have met the minimum training requirements, but will also give a focus for lines of questioning during the final assessment interview. It is therefore vital that you manage and record this information during the training period in an organized and systematic manner, as this will greatly assist you in the final interview.

THE RECORDS

Most of the records you need to keep have been considered briefly in outline in the previous chapters. The following guidance pulls all of this together and provides some practical tips and advice.

Diary

A diary must be kept by all candidates. This is simply a day-to-day record of your training and experience. The cover page comprises a record of your name, date of registration for APC, route, supervisor and counsellor, and counsellor's declaration confirming that the diary and log book are a true and accurate record of your work. Figure 4 shows a typical format for the diary pages.

Week ....

Date	**Nature of professional work carried out**	**Competency reference**

(Further details can be found in the templates at the back of the *candidate's guide*.)

Figure 4 Typical format for diary pages

The diary will provide the information and detail that you will need to include in your interim and final records. It will illustrate and bring to life details in your other documents and thereby help the assessment panel to target their questions.

With the exception of the Minerals and Environmental Management and Marine Resource Management routes, your diary does not have to be submitted as part of the final assessment documentation. However, it should always be kept up to date and be available for inspection by your RICS training adviser (RTA) on an office visit or by the final assessment panel, if requested. There may be occasions when the assessment panel, through RICS, may call for your diary to be submitted, to form a more detailed view of your training period.

You can only start recording experience after you have received confirmation of your enrolment from RICS. The date shown in the acknowledgement of enrolment from RICS will be the first entry in your diary. Remember that if you change employer, this must be clearly marked, with perhaps a couple of parallel lines to show a break and a few words of explanation (see figure 5). You must also advise RICS of this change.

CHANGE OF EMPLOYER	25/9/2001
NEW EMPLOYER	Smith & Co Chartered Surveyors 35 High Street Birmingham B1 1AA

Figure 5 Diary entry indicating a change of employer

There are four important pieces of advice regarding your diary:

1 keep it up to date: do not let the weeks slip by without completion. Get into the habit of writing it on a daily basis. Can you remember how you spent the day, say, two days ago? It is likely that you will have forgotten more than you can remember;
2 get into the habit of using descriptions that relate to the various competencies, so that you can make easy links to your record of progress and competency achievement planner. For example, 'negotiation and agreement of repair works under a final schedule of dilapidations' will give you all of the links and reminders you need to properly complete your other documentation, such as the competency monitoring table. It will also help you when you are reviewing your overall performance against your competency achievement planner;
3 you should use your diary as part of the preparation process for the final assessment. It is therefore important that it is well

written and clearly presented, to help you when you are reviewing your training and experience over the two-year period, in preparation for the final assessment interview;

4 keep a note of training and experience attained in respect of the mandatory competencies as either a full half-day or quarter-day, if appropriate, or simply a brief note as a reminder, possibly with a number of hours, so that this experience can be transferred to your log book, professional development record and record of progress.

Log book

The log book is a summary of the training recorded in your diary, on each competency, within every 12 months of the training period. Each page will show your name and route and a simple record, as indicated in figure 6.

	Months												
Competency title and number	1	2	3	4	5	6	7	8	9	10	11	12	Total number of days

Figure 6 Format for the log book

Get into the discipline of completing your log book at regular intervals, both at the end of each month and at the end of each 12 months of training, as part of the interim assessment. The information in the log book will be sent to RICS as part of the final assessment submission to be used by the assessment panel.

The log book is a very useful tool for the assessment panel. It provides an immediate snapshot of your areas of work experience and will be used, in conjunction with your record of progress, to structure the final assessment interview in order to obtain the correct balance of questioning relative to your experience.

Professional development

Be proactive in recording your professional development. As explained in chapter 2, there are many aspects of your daily work that count as professional development, such as attending meetings, preparing to run a meeting or give a talk, structured reading, and so on. Make sure that you are fully aware of what constitutes professional development, and keep a regular record in your diary. This record can then be transferred on a monthly basis to your log book.

Record of progress

The record of progress templates at the back of the *candidate's guide* enable you to keep a record of your development against the competencies. These forms comprise the three-monthly supervisor's reports; the six-monthly counsellor's reports; the interim and final assessment records; and the referred candidates' form. There is a cover page similar to that for the diary, with the exception of the declarations. The declarations are made by your supervisor in addition to your counsellor, confirming that the minimum competency requirements of the APC have been achieved. They must be completed for both the interim and final assessments.

There are two formats for recording progress: one for the mandatory competencies, which simply replicates the statement from the appropriate level in *APC/ATC requirements and competencies* and provides margins for the supervisor's and counsellor's signatures; the other specifically designed for the core and optional competencies, which starts with a column for the reference number of the particular competency. This reference number may be found in brackets at the end of each competency statement in part two of *APC/ATC requirements and competencies*. There is a column for the title of the competency and then margins for the supervisor's and counsellor's signatures against each level of attainment. A typical example can be seen in figure 7.

Number	Title	Level	Supervisor	Counsellor
69	Valuation	1	*SJ 02/01/01*	*JW 10/01/01*

Figure 7 Typical entry in the record of progress

You should aim to make the record of progress the focal point of your meetings with your supervisor at three-monthly intervals and with your counsellor at six-monthly intervals. You need to be proactive in managing your progress. Therefore, in advance of these meetings, prepare a note of how you feel your training is going, how you are developing against the competencies and how far you think you have progressed against the various levels for the mandatory, core and optional competencies. You can then compare notes with your supervisor and counsellor.

Template 6 in the *candidate's guide* does not show the mandatory competencies in levels, as with the core and optional competencies. My advice is that your supervisor and counsellor should add a note in the margin showing the level that is being signed off.

Progress reports

The role of your supervisor and counsellor throughout the training period is most important. Detailed guidance on these roles and responsibilities has been included in the *APC guide for supervisors, counsellors and employers*. I do not intend replicating this guide. However, there are a few key issues worth noting:

- the supervisor is responsible for overseeing your day-to-day work, whereas the counsellor is responsible for managing your training at a strategic level;

- ideally, the supervisor and counsellor should be different people. However, the roles may be combined if, for instance, you are employed by a sole practitioner;
- ideally, both parties should be chartered surveyors. However, the supervisor may be a member of some other professional institution. If this is the case, the counsellor must be a chartered surveyor;
- the supervisor should give guidance, support and encouragement on a daily basis. At three-monthly intervals, he or she will assess your progress against the competencies and complete your progress reports;
- the counsellor has a similar role, and will formally assess your progress on a six-monthly basis and complete the appropriate progress report. In reviewing your overall progress, the counsellor will provide a second opinion to that of the supervisor;
- your supervisor and counsellor have responsibility with regard to confirming the declarations on the front page of your diary (counsellor only) and your record of progress (supervisor and counsellor);
- at the interim assessment, which should be completed within one month of the first 12 months of training, your supervisor and counsellor will be responsible for certifying that the diary, log book, record of progress and interim summary of training are true and accurate records of your training and experience to date;
- prior to the final assessment, your supervisor and counsellor must certify that the log book, diary (for Minerals and Environmental Management and Marine Resource Management candidates only), final assessment record, record of progress and critical analysis are true and accurate records of your training and experience.

You must take a positive approach to the meetings during the training period to discuss progress. Make sure that you agree the times and dates for these meetings and are prepared to give your input. As explained previously, use the record of progress as the focal point for these meetings. Remember that your final assessment will be based on your performance and competency statements and levels, and it is therefore important that your training and experience are geared towards this. The regular meetings with your supervisor and counsellor will be the ideal opportunity to review progress and to forward plan how you will fill any gaps.

It is also worth noting that in mid-2000 RICS introduced a system to monitor the accuracy of supervisor's and counsellor's declarations. Chairmen of assessment panels have been given authority to request that RTAs visit employers where doubts have been raised concerning candidates' readiness for the final assessment.

INTERIM ASSESSMENT

Within one month of recording 12 months of training, or on the completion of your sandwich placement, you should, in conjunction with your supervisor and counsellor, complete an interim assessment of the training and experience that you have gained up to that date. The date of the interim assessment is important, because a further 12 months of training must be completed before you can apply for the final assessment.

This is a highly significant juncture of your APC, giving you the opportunity both to review progress and to plan the final 12 months of training. There are three forms that need to be completed:

1 progress to date: this form is based on the monitoring tables which you will have used in conjunction with your competency achievement planner. It requires you to write, in approximately 1,000 words, an account of your first 12 months' training and experience. You can draw upon the day-to-day detail noted in your diary to complete this. There is also a column in the form that allows you to plan your training for the second part of your training period. This may comprise a note of competencies or levels where further experience is needed;

2 forward plan: this form requires you to write, again in approximately 1,000 words, an explanation of how you will gain the training and experience that is necessary in the second part of the training period. It should be written with regard to the competency statements and levels;
3 supervisor's and counsellor's report: this form will mainly be completed by your supervisor and counsellor. It will draw together the information contained in your supervisor's three-monthly reports and your counsellor's six-monthly reports. There is also provision for your comments. The form contains a section for certification by all parties declaring that the interim assessment has been completed and that your diary, log book and record of progress have been correctly completed and maintained.

You should also include in the forms some information about your professional development. This should comprise a history of the main elements of your programme of structured professional development, together with a summary of the knowledge gained from each. As explained in chapter 2, do not forget that at both the interim and final assessments, the assessors will be looking for clearly defined links between your professional development and the competencies. There must be evidence of a planned and systematic approach, as opposed to an *ad hoc* or random selection of professional development topics. For example, if you are training to be a valuation surveyor, a two-hour lecture on rocket science would not be an appropriate choice of professional development.

The layout, content and detail on the various forms will steer the discussions with your supervisor and counsellor. All of the forms can be found at the back of the *candidate's guide*, along with comprehensive guidance on completing them. The forms do not need to be submitted until the final assessment, when they will form part of the final assessment judgement of the panel. You should also be aware that your RTA may wish to see these documents. Failure to provide evidence that the interim assessment requirements have been completed may result in your final assessment being delayed. You should therefore make sure that the interim assessment is completed on time and is readily available.

The best advice I can give you is to use the interim assessment to help the final assessment panel to target questions to you about your actual experience. When completing the 'summary of experience/training completed' part of the template, fill it with examples of the work that you have been doing to gain experience in the various competencies. Tell the assessor about the projects, inspectors, valuations, issues and problems that you have encountered and dealt with in your day-to-day experience. This will give the panel the lead-ins they need to start questioning you in the final assessment interview.

SUMMARY

- The information and evidence contained in your diary, log book, professional development record and record of progress are very important, as they will be used in the final assessment to satisfy the panel that you meet the training and experience requirements of the APC.
- Your diary must be available for inspection if requested by the RTA or the assessment panel.
- Get into the habit of completing your diary daily and your other records at regular intervals.
- The record of progress comprises a series of forms to show your development against the competencies.
- Be proactive in all meetings with your supervisor and counsellor. Plan and prepare for them.
- Remember that in the interim assessment you must provide evidence of a planned and systematic approach to professional development.

Preparation for the final assessment

This chapter covers the months in the run-up to the final assessment interview. It considers the paperwork that needs to be completed for the final assessment. In order to assist your preparation, it provides detailed guidance on the methodology that will be adopted by the panel during the interview. Finally, advice on preparing the final assessment record and the critical analysis is given.

THE PAPERWORK

You will recall from chapter 1 that in the letter you receive from RICS confirming your enrolment, you will also be given a likely date for the final assessment. This date will be held on record by RICS and five months before the final assessment dates, you will be sent a final assessment application pack. There are two important points to note with regard to this:

1. if you do not wish to sit your final assessment on the proposed date, the onus is on you to notify RICS that you are deferring;
2. if you have any special needs or disabilities, RICS will, subject to notification, take appropriate measures to assist you at the assessment centre. There is a box on the application form to be completed if this is the case.

After receiving your final assessment application pack, you must send in the completed application form. You then have approximately one month to complete and send all of the other required documents to RICS. With your application pack you will receive guidance on completing the various forms, as well as a useful checklist comprising a list of the documents that need to be forwarded to RICS. There is a checklist for candidates applying for the first time and another for candidates who have been previously referred.

You must complete part of the front page of the assessment panel's marking sheet with personal details and a passport-sized photograph. You must also complete the final assessment application form with various personal details, along with information concerning your training and experience. The latter is very important, as it will assist RICS in putting your final assessment panel together. RICS will match panels and candidates in terms of background, training and experience, so that the detail of the interview may be correctly targeted by assessors with experience in similar areas to your own. The other main form that you must complete as part of the application pack concerns your professional education and employment details, outlining your history of education and employment up to the date of the final assessment.

When applying for the final assessment, you may not yet have completed the minimum 24 calendar months of training and therefore may not have reached the required levels in some of your competencies. If this is the case, and you do not wish to defer, you must complete a declaration stating that at the date of the final assessment you will have reached the required levels. The assessment panel will seek to confirm this on the day of your final assessment.

Don't forget, if you need any help or advice, you can contact RICS.

THE FINAL ASSESSMENT INTERVIEW

This book has been written in a logical order, that is to say, following the actual course of your APC. At first sight, therefore, it may appear that consideration of the final assessment interview at this juncture is a little premature. However, the reason for this is to give you an understanding of competency-based interviewing, so that you can start the mental preparation for the interview at an early stage. In particular, it is important that you are focused on this concept in the two-three month run-up period in which you will prepare your critical analysis and presentation.

I would also like to add a personal observation. I have been involved with APC training for a decade, and have not yet met a candidate who has been able to turn up on the day without any preparation or revision and be successful. I can only compare the final assessment to training for a major sporting event. To be successful, an athlete needs to approach the run-up to the big event in a planned and structured way, looking to peak, in performance terms, on the day. You will need to take the same approach to the run-up to your final assessment. Put the time and effort in, and you will be more likely to succeed.

In the lead-up to the big event, you need to have a clear understanding of competency-based interviewing, so that you can take an informed approach to the revision and preparation needed for the final assessment interview.

All good interviews essentially have eight important aspects:

1 objective;
2 criteria;
3 role of the chairman;
4 structure;
5 questioning technique;

6 note-taking;
7 equal opportunities;
8 conduct, best practice and customer care.

This chapter will look at the objective and criteria in terms of how they link to competency-based interviewing. I will deal with the remaining aspects of the interview in chapter 5.

The objective

The objective of any interview must be clearly defined at the outset. Interviews are conducted for a variety of reasons – for employment, promotion, appraisal or dismissal.

The objective of the APC is to ensure that only those candidates who have an acceptable level of competence in carrying out the work of a professionally qualified surveyor on behalf of a client or an employer are admitted to professional membership of RICS. This objective is set out in the first part of the *candidate's guide* and is simply designed to test whether you are 'competent to practise' as a chartered surveyor.

Your competence will mainly be judged by the assessment panel asking questions, many of which will test your ability to put theory into practice. Throughout the interview, the assessors will set questions based on everyday problems faced by practising surveyors. This is an important concept. The APC is a *practical* test and assessment panels are made up of practising surveyors who are best placed to test your competence – based on their own up-to-date knowledge and current experience of the problems, issues and difficulties faced by the profession.

The criteria

In any interview situation, it is important that criteria are set. The criteria are the standards or benchmarks against which all candidates will be judged. They provide the consistency and uniformity required to create fairness, or equality of opportunity, for all candidates. For the APC, the criteria are set out in the front of the *candidate's guide*. They can be divided between the immediate criteria that the panel will apply on the day, and the developmental criteria which have a longer-term connotation. Management is a good example of the latter. Management skills and abilities, if nurtured and developed, evolve over a period of time, whereas certain aspects of technical development can occur more immediately.

The immediate criteria that will be tested by the panel are that you:

- 'have learnt to apply your theoretical knowledge through professional training and experience to attain practical skills;
- have achieved a satisfactory level of understanding and application of the skills that form an essential part of the knowledge base of your chosen route;
- are aware of the need to pay particular attention to accuracy and essential detail to safeguard the interests of employers and clients;
- can communicate effectively – orally, in writing and graphically – and prepare reports which are well structured, grammatical and spelt correctly;
- are aware of and intend to act in accordance with the Rules of Conduct; possess the highest level of professional integrity and objectivity; and recognize your duties to clients, employers and the community.'

With regard to the developmental criteria, you will be expected to demonstrate that you have developed so that you:

- 'are a good ambassador for your profession, the RICS and your employer;
- are aware of the professional and commercial implications of your work;
- understand your clients' and employer's thinking and objectives;

- have an up-to-date and developing knowledge of legal and technical matters relevant to the work that you do and the law of the region or country in which you practise;
- are able to play a role in a team and build up experience in client contact;
- are aware of the operation of general economic principles;
- have developed the confidence to work unsupervised;
- are able to demonstrate motivation, initiative, and administrative ability.'

These criteria overarch the whole assessment process. If you think of the objective – becoming competent to practice – as a mission statement, the criteria can be seen as the detail that underpin that statement. Consider, for example, what 'competent to practise' means. In the first instance, it means having learnt, through training and experience, to apply your theoretical knowledge to practical situations – the first of the aforementioned criteria.

Additional criteria have been set for the critical analysis and the presentation, which I shall deal with in this chapter (the critical analysis) and chapter 5 (the presentation).

The part competency-based interview

The final assessment comprises a part competency-based interview. The objective of a competency-based interview is to allow you to demonstrate your skills and abilities against the various competencies that form the requirements of the training period. The assessors will ask questions and set oral problems that draw on your experience and allow you to demonstrate your problem-solving abilities. Whereas in a criterion-based interview the interviewer would mainly focus on knowledge and skills, in a competency-based interview, he or she will also consider attitude and behaviours. The philosophy behind the competency-based interview is that past behaviour is the best predictor of future behaviour.

You should note that the final assessment is a *part* competency-based, as opposed to *full* competency-based, interview. In a full competency-based interview, the interviewer would not question you outside of your actual training and experience. However, the objective of the APC is to consider whether you are competent to practise as a chartered surveyor in the widest sense, and the assessors will therefore test knowledge of wider issues, as well as seeking to ascertain your awareness of your limitations and your knowledge of matters of wider concern to the profession. On occasions, therefore, you will step outside of your actual training and experience. A classic example of this relates to the Rules of Conduct. The senior partner or managing director of your firm will usually deal with many aspects of the Rules of Conduct, such as the client account regulations, and in normal circumstances you will not gain hands-on experience of such matters. However, you will still be expected to have some knowledge of the basic issues.

The assessors may also wish to test your ability to take learning and experience in one area and apply it elsewhere. That is, they may test skills transfer – stepping outside of your actual training and experience to ask you to apply what you have learnt to a hypothetical situation.

So how will the objective, criteria and competencies be drawn together by the assessment panel during the course of the interview? The answer is quite simple. The panel will take a competency and test a variety of the criteria against this, with a view to addressing the objective of the interview: are you competent to practise as a chartered surveyor? For an example of this, see figure 8.

The aim will be to test you over the full range of the criteria during the course of the interview. However, it is not necessary to test all the criteria against every competency. The testing of the criterion relating to your ability to put theory into practice, for instance, will also test a number of other criteria. In the example on page 49, the test of your ability to put theory into practice will also serve to demonstrate your knowledge of legal and technical matters, operation of general economic principles, and so on. Generally, you will find that in the course of testing your ability to put theory into practice, an experienced assessor will draw out details of your knowledge, skills and abilities in relation to a large number of other criteria.

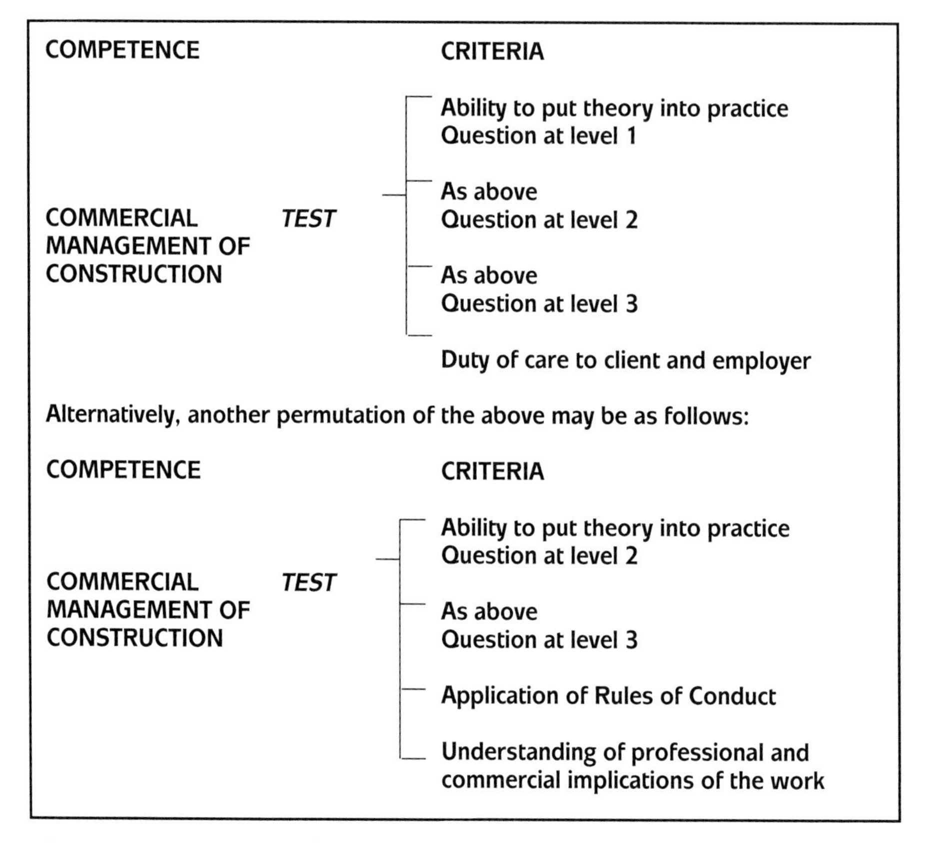

Figure 8 Questioning for competence in the construction faculty (quantity surveying)

An example of questioning in line with figure 8 might be as follows:

- 'Having regard to your experience and the projects with which you have been involved, what were the main criteria considered in deciding upon a method of procurement?' (level 1);
- 'I note from your final record of experience that you have dealt with a design and build contract involving a 1-hectare greenfield site, upon which was constructed a very basic 5,000m office block over a six-month time period – to a very basic specification. How did you advise that client over the course of the project? What were some of the key stages/issues? What about cash flow/variances? (level 2);
- 'How did you go about preparing your final reports to the client? What issues did you include?' (level 3);
- you might also be questioned on professional indemnity insurance requirements, thereby touching on the Rules of Conduct;
- this could be followed by an assessor asking you, 'during the course of this contract, what quality control and assurance measures did you put in place to protect your client's interest?' (covering duty of care);
- finally, you might be asked for 'a broad indication of the construction costs per m^2 for the project', in order to test your commercial awareness and knowledge of the professional implications of your work.

At the end of the interview the assessment panel will make a judgement as to whether you have demonstrated the required level of competence. As shown in the example above, the panel will do this by taking any of the appropriate competencies and asking questions to ensure that you meet the criteria that have been set. On this basis you will then be considered for professional membership of RICS, as the overall objective of the APC will have been met.

It is also important to note that in a 60-minute interview the panel will not be able to test a full two years of training and experience. They will need to be selective – so don't worry if there are gaps and omissions. The panel may have been satisfied by some other piece of evidence, in the written submissions, concerning your competence: it is not all down to questions and answers in the interview.

Having explained the basic concepts of a part competency-based interview, what follows is guidance on how you should build this into your revision and preparation for the final assessment. See the examples below for further advice on the times when the interviewer will step outside of your actual training and experience.

- One candidate told me that he knew nothing about the Rules of Conduct because the senior partner in the firm dealt with these matters. However, even in the absence of hands-on experience, the criteria set for the final assessment expect you to have some knowledge of the Rules of Conduct.
- Consider yourself as a quantity surveyor having only hands-on experience of small works types of contract. This can happen, but you will still need to have an understanding of other types of contract.
- Think about politics and economics and the effect that government economic policy can have on interest rates. These in turn can affect all aspects of the property world, the construction industry and so on.
- Some years ago I observed an interview and felt that an assessor had been hard on a residential property candidate. The candidate had no experience of council

tax, but a number of questions were asked on this. The assessor's reply was that he could stop someone in the street and most likely get the answers to the questions asked.

- On another occasion, I watched a valuation surveyor with no experience in rating being asked some very basic rating questions. He was a commercial management candidate and the scenario was one of questions being asked by a tenant of a shop about a large rates demand. The assessor's view was that he should know a little about the subject, if nothing else, to understand his limitations and advise the tenant where to seek advice.

These examples should give you an understanding of why the final assessment needs to be a part competency-based interview. They should also provide some guidance in terms of the background reading you will need to do in preparation for the interview. Sometimes, when I am training candidates in preparation for the final assessment, I run a brainstorming exercise. I ask everyone in the group to prepare a list of background preparatory reading for the final assessment – we often fill up to 10 pages of a flip chart. It is an exercise that I would strongly recommend you carry out while planning your preparation and revision.

The remainder of this chapter will consider two other aspects of your preparation and run-up to the final assessment: the final assessment record and critical analysis.

THE FINAL ASSESSMENT RECORD

The format of the final assessment record is similar to that of the interim assessment outlined in the previous chapter. However, for this summary you will only use the progress to date and the supervisor's and counsellor's report forms. The objective is to give the assessment panel an outline of your training during the 12 months prior to the final assessment. This should be completed in approximately 1,000 words and should be based on your progress against the competencies. As with the interim assessment record, my advice is to use this as an opportunity to give the final assessment panel lots of practical examples of the work that you have been doing, to help them to properly target their questions to your training and experience.

In the progress to date form, the column entitled 'training planned' may be used to outline experience to be gained in the final three months before the assessment. The reason for this is that there will usually be a period of three months between submission of the documentation and the final assessment. During this period it may be that you plan to acquire further training and experience in some of the competencies. This information will assist the panel in forming a complete picture of your training.

The supervisor's and counsellor's report form will draw together the information contained in your supervisor's and counsellor's progress reports for the second 12 months of the training period. It will include your comments and provide space for certification that the final assessment records have been completed and that your diary, log book and record of progress have been correctly completed and maintained during the second part of the training period.

THE CRITICAL ANALYSIS

The critical analysis is a written report of around 3,000 words. It is a detailed analysis of a project with which you have been extensively involved during your training period, with a conclusion giving a critical appraisal of the project, together with a reflective analysis of the experience gained.

In recent years the choice of topic for the critical analysis has been the source of much confusion. You may be working for a large firm and have been involved with an instruction or project that is considerable in size or importance. Your role in the instruction or project would therefore be an appropriate topic for the critical analysis (see the further guidance on page 54 concerning 'key issues'). On the other hand, the instruction or project that you choose need not be particularly complicated, or of great value. It may simply be typical of the type of work with which you have been involved during your training period.

It must be emphasized that you are not expected to be running the project you choose. It is your involvement or role in the team that you are expected to outline, analyse and provide comment on. In addition, it is not necessary that the project has a definite start and finish point. It may be that at the time of writing your critical analysis, the instruction or project has not reached a conclusion. Your report will comprise the detail up to the date of writing, and perhaps contain a prognosis of the outcomes. If the outcomes are known at the date of the final assessment, you may wish to include them in your presentation.

The report should be a maximum of 3,000 words and supported by an appropriate number of appendices. However, it is quality not quantity that is important, so use a word count and do not include too many, or too lengthy, appendices. Appendices should be included to support your report, not to add or expand upon it.

The format of the critical analysis is clearly set out in the *candidate's guide* and it is important that this is followed. One of the main reasons for referral is that this guidance has not been followed and that the format of the critical analysis is too similar to the type of report that you would write at work. The main headings of the report should be as follows:

- key issues;
- options;
- reasons for rejecting certain options;
- proposed solution to the problem/s and reasons for this choice;
- critical appraisal of the outcome and reflective analysis of experience gained.

The assessors are also looking for good communication skills, so think about the layout of the report, the presentation and use of photographs and plans, the grammar, spelling, number of words, index, pagination, and so on. Put yourself in the position of a potential client picking the report up for the first time. Would you be impressed with the presentation, the content and the advice? Would you consider awarding the contract?

As the assessors are also looking for a high level of professional and technical skills, make sure that this aspect of your report is checked and double-checked.

Now let us consider each of the above headings in turn.

Key issues

The project that you have been involved in could be fairly extensive. If you select too many key issues, it is likely that you will merely skim the surface of them, rather than undertaking a detailed analysis. So be selective. You may wish to select just one key issue. It is possible that this issue will be common across a number of projects with which you have been involved, and this is acceptable. Overall, to meet the requirement of a 'detailed analysis', think about the depth required as being about level 3 of the various competencies involved.

Options and reasons for rejecting solutions

It is uncertainty that creates the need for experts, and it is the diversity of solutions to any particular problem that leads clients to request professional advice. Therefore, before proposing a solution to a client, you will need to consider all of the options. Under this heading, you must demonstrate your ability to think laterally and must show that you have genuinely considered other options to your preferred solution. Give reasons as to why some solutions were not feasible. Do not fall into the trap of going down one route only. The guidance in the *candidate's guide* clearly requests that you consider various options or possible courses of action and also that you give reasons for the rejection of those options not selected.

Your proposed solution

You must give a detailed account of the reasons supporting your adopted course of action. Here again it is important that your thoughts cover a broad canvas. Too many reports simply deal with the technical aspects of a particular job. Think about all the aspects that support your decision: financial, technical and professional, and issues relating to customer care, the Rules of Conduct, ethics and conflicts of interest.

Critical appraisal and reflective analysis of experience gained

The conclusion to your report must include a critical appraisal of the outcomes, together with your own views and feelings as to what you have learnt from the experience. This part of your report may account for around one-quarter to one-third of the total number of words.

The critical appraisal is all about introspection. All good professionals need to be able to look at a project, consider what they have done well, identify what they have not done so well, and plan how they might improve upon their actions the next time they carry out a similar task. Consideration of all of this will comprise your critical appraisal of the project.

The next step is to stand back from the project and reflect upon what you have learnt from the experience gained.

The assessors will take your critical appraisal as a starting point to question you beyond what you actually did and to probe your understanding of the wider issues surrounding your project. It is therefore useful to start thinking about these aspects while you are writing your critical analysis, rather than waiting until you are in the interview.

Sadly, these last two aspects of the critical analysis have proved to be the most lacking in recent years. There have been more referrals as a result of these aspects than for any other reason.

The critical analysis is a professional piece of work and should therefore be signed and dated by you. Don't forget your supervisor's and counsellor's certification, as required in the *candidate's guide*.

One final piece of advice. Over the years I have been approached by many candidates concerning confidentiality. I can assure you that the information contained in your critical analysis will be treated in the strictest confidence by the panel.

REPORT WRITING

I would like to close this chapter by giving you some practical guidance on report writing. With regard to the critical analysis, if you follow the 10-point plan set out below, together with the practical guidance covered above, you should be well on the way to success:

1 the objective of the report: never lose sight of it! In the critical analysis, you are looking to write a detailed analysis of a project (or projects) with which you have been involved during your training period. The conclusion to this report involves a critical appraisal of the outcomes, together with a reflective analysis of the experience gained. Keep this in mind throughout;
2 brainstorm the subject: brainstorming is a simple technique where you take a topic or subject and write notes, in no particular order, of all of the thoughts and considerations that come to mind;
3 prepare an outline and consider the appendices: start to put your brainstorming into some order, using headings and important issues. Then consider the appendices that will be useful to support or shed light on the critical analysis. Do not fall into the trap of using the appendices to add to the volume of words; it is quality, not quantity, that the assessors will be looking for;
4 consider the use of visual aids, plans and photographs: for simple reasons, photographs can speak a thousand words and you have been set a limit of 3,000 words for the text! But do think carefully about the need for and use of any visual aids;
5 stand back and review where you have got to: are you still on course to meet your objective? Does everything look interesting and entertaining? How would you view the content if you were on the assessment panel?
6 write the report out in full: start pulling the detail together, in terms of the text and appendices, plans, photographs, and so on;
7 once again review the report: look at the number of words, use a word count, and be particularly careful with spelling and grammar;
8 polish the report: think of an attractive cover page and give some thought to the index, binding and paragraph numbering. Make sure that the report is signed and dated and certified by your supervisor and counsellor;
9 test the report for potential areas of questioning: don't forget, the panel will be extending their questioning beyond what you actually did, and will also probe your understanding of any wider issues surrounding the project. It might be useful to ask one or

two of your colleagues in the office to read your critical analysis, with a view to asking you some questions along these lines.

10 think about the objective once again: has it been met? Go back to where you started and consider the following:
    - have the key issues been made clear?
    - have you considered the options and, in particular, the reasons for rejecting certain options/solutions?
    - is the proposed solution supported by detailed reasoning?
    - does the conclusion include a critical appraisal of the outcome and a reflective analysis of the experience gained?

The critical analysis is not an examination and you will have access to texts, references and a whole host of technical and professional references when writing it. It is therefore important to narrow the key issues down, so that you can write to the appropriate level of detail. In doing this, take care to ensure that the technical and professional references are to a high standard, as this will be an important issue in the eyes of the assessment panel.

CRITERIA TO BE APPLIED BY PANEL

In summary, the criteria the assessors will be looking to apply to the critical analysis to decide whether it should be passed or referred are as follows:

- have the key issues been identified?
- have all the options been considered?
- are the reasons for rejection of certain options clearly stated?
- is the preferred solution supported by sound judgement?
- does the conclusion contain a critical appraisal and reflective analysis of the experience?
- has the candidate demonstrated high standards of spelling and grammar?
- does the report indicate high standards of technical and professional skills?

You will recall that earlier in this chapter the criteria that 'overarch' the final assessment process were considered. The above criteria clearly set out what the assessors will be additionally looking for in the critical analysis.

SUMMARY

- Approximately five months before the final assessment date you will receive an application pack from RICS.
- You must advise RICS of any special needs or requirements for the final assessment interview.
- Make sure that you follow the guidance that accompanies the application pack. Remember to use the checklist provided by RICS.
- The final assessment interview is part competency-based, so do not lose sight of the wider issues.
- Always stay clearly focused on the objective of the interview and on how the competencies and criteria fit together.
- The final assessment record is an important document, as it will help the assessors to target questions and lines of inquiry. Give plenty of examples of your work and make sure they are a true and accurate reflection of what you have been doing during the training period.
- Ensure that the format of your critical analysis complies with the guidelines set by RICS.
- Ensure that the critical analysis addresses the key issues of the project you have selected and gives reasoned support for your chosen solution or proposal.
- Give proper consideration to the critical appraisal and reflective analysis that is required in the conclusion to the critical analysis.
- Spelling, grammar, presentation and a demonstration of a high standard of technical and professional skills are also very important aspects of the critical analysis.
- Most importantly, ensure that your critical analysis meets the criteria that the panel will apply when deciding whether to pass or refer.

The interview and presentation

This chapter sets out the standard components of the final assessment interview. It seeks to expand and develop your knowledge and awareness of the role of the interviewer, and in so doing, to develop your technique as an interviewee. The presentation is also considered in detail.

COMPONENTS OF THE INTERVIEW

Your final assessment interview will contain a number of standard components. After the welcome by the chairman, your first task will be to give your presentation to the panel. The panel will then question you on the issues raised by this. Afterwards, they will question you on your wider training and experience, with relation to your chosen competencies.

Candidates for the Antiques and Fine Arts route will, in addition, be required to undertake a written valuation on 15–20 objects, provided on the day, for insurance purposes. This will be followed by a discussion on the valuation and then an inspection of a range of objects, which will be handled, enabling candidates to demonstrate how they would advise clients verbally. I strongly recommend that candidates for this route obtain further details of these additional requirements from the RICS Practice Qualifications Department.

OVERVIEW OF THE INTERVIEW

The interview can be described as an information-gathering process in which the interviewer's most important skill is questioning technique. You will do the majority of the talking, with the split in this respect ideally in the ratio of 70:30. At the end of the interview, the assessment panel will consider the information gathered and how well it meets the criteria set for the interview. This in turn will determine whether the objective has been met, which will then determine the outcome: have you been successful?

In chapter 4 we considered the various aspects of the interview, concentrating in particular on the objective and the criteria. I now want to look at the other aspects in more detail:

- role of the chairman;
- structure of the interview (including the presentation);
- questioning technique;
- note-taking;
- equal opportunities;
- conduct, best practice and customer care.

Under each heading below, an explanation of how the panel will operate is provided, followed by guidance on how you can improve and enhance your technique, with a view to delivering an effective performance – i.e., a performance that persuades the panel that you are 'competent to practise'.

You should appreciate that every interview you face will be different. In the context of the APC, no two assessment panels will operate in exactly the same manner. Therefore, the information that follows is provided in the context of guidance only. However, it is hoped that it will be helpful to you.

THE ROLE OF THE CHAIRMAN

The chairman of the assessment panel has a very important role to play during the interview. Let us now look at this role in more detail, with particular regard to the various aspects of the chairman's performance that are aimed at assisting you.

The chairman will have made contact with the other panel members shortly after receiving your detailed information and final submissions from RICS. He or she will have held a brief discussion with the panel members concerning various aspects of the interview and will usually have arranged to meet them one hour before the first interview.

A critical part of the interview is the opening three to four minutes. During this period the chairman's conduct will be geared towards settling your nerves. The final assessment interview can be considered as an 'advanced role play situation'. Consequently, you will probably be more nervous than you would be in a job interview. You could argue that there is much more resting on the outcome of this interview, with many years of effort having gone into the preparation, than on any job interview. Therefore, the importance of settling your nerves and 'breaking the ice' is vital.

Figure 9 is a simple stress management graph that explains this concept in more detail.

At the outset of the interview your nerves will be high. The chairman and panel members will aim to help you settle these nerves. Just before the end of the interview, when the chairman signals the close, your nerves will rise again for one last time. Your nerves may also increase somewhat as the interview progresses and you move from one panel member to another.

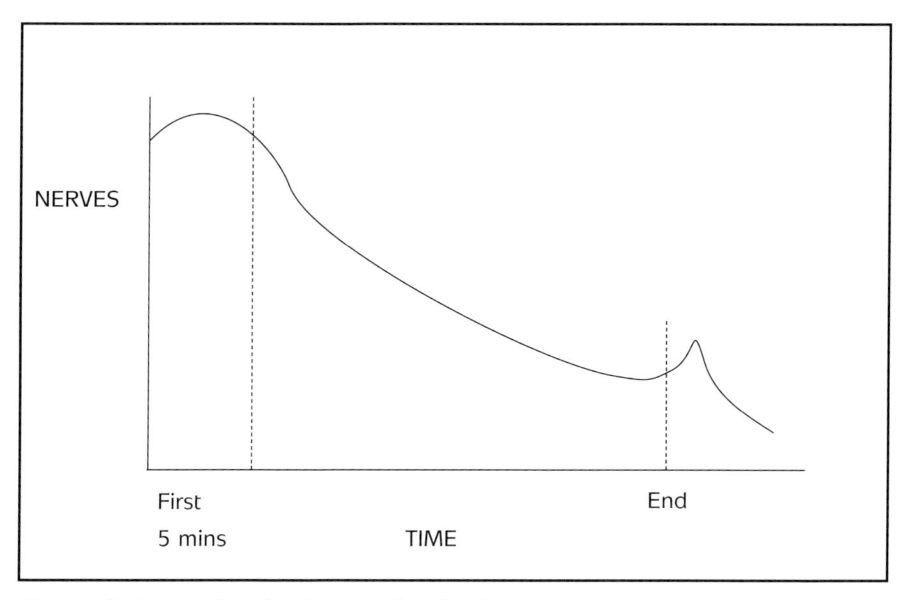

Figure 9 Stress levels during the final assessment interview

So how will the panel try to help you settle your nerves? There are a number of things they can do:

- introductions: as soon as you enter the room, the panel will follow an agreed format in terms of handshakes and the provision of names and any relevant background information;
- structure: the chairman will give a brief outline of the structure of the interview. This will give you an idea of how the panel intends to organize the time and will help you know what will happen next;
- comfort: the panel will make you feel comfortable by offering you a seat, water, the opportunity to take your jacket off, and so on.
- notes: the chairman will explain that the panel will be keeping notes. Do not be put off, therefore, if you see the panel scribbling away furiously. This is done to assist with your assessment at the end of the interview;
- health check: after you have settled in, and before you move on to the presentation, the chairman will check to ensure that you

are 'fit and well and ready to proceed' – or similar words to that effect. This check is to ensure that all candidates are given equality of opportunity. If you feel unwell before the interview, it is important that you do not proceed. You should tell the chairman, who will advise you what to do
- the ice-breaker: the opening question asked by the chairman will be designed to 'break the ice'. The only way that you can get rid of nerves in an interview is by talking them out! The chairman will ensure that your first question is easy to answer and may be phrased something along the lines of, 'Tell us in your own words what sort of work you have been involved in over this last six months'. Make sure that you are prepared for this question. Practice your response in advance, as this will help you at the beginning of the interview, when your nerves will be at a peak;
- last word: the panel will always let you have the last word. The chairman will say something like, 'At the end of the interview I will give you the opportunity to come back on, add to or clarify anything that we have discussed'.

The chairman may also provide the following additional support:

- have a word with you outside the room, run through the interview structure briefly and then escort you into the room to meet the panel;
- keep you briefed at each stage of events, even though the structure has been outlined at the outset. This will help you with your nerves and will be designed to give the interview some sense of continuity. The chairman may use phrases such as, 'Thank you for your presentation, we are now going to discuss some aspects of it with you. My colleague … will begin'.

Your role

There are a number of ways that you can help yourself prior to the interview:

- make sure that you know where the assessment centre is. Time your journey to give yourself plenty of time. Try not to put pressure on yourself by worrying about these things on the day – sort them out beforehand;
- if you have had any problems with the journey, or are unwell on the day, let the RICS staff know. This can be taken into account during the interview;
- look the part. There is no doubt in my mind that a good first impression will give you a mental boost and will help with your nerves and confidence;
- rehearse your opening lines and prepare for the ice-breaker.

First impressions

Figure 10 shows how we let first impressions, which are often made in the first few seconds of meeting someone, form our views and opinions.

When we meet someone for the first time, our initial impressions will be based on such things as stature, dress, deportment, etc. When the person speaks, they add to the mental picture we are building. We may immediately relate to the person's dress sense or accent, or may find that these do not conform with our 'norms'. We may therefore create a 'barrier', which can affect judgement and prejudice our views and opinions. It is only when we get beyond these first impressions that we really get to know someone and develop a deeper understanding of them.

I should point out that the assessment panel will have been trained to deal with first impressions. To counterbalance the effect of such impressions, they will probably take notes during the interview, to ensure that you are judged on your overall performance, rather than initial impressions.

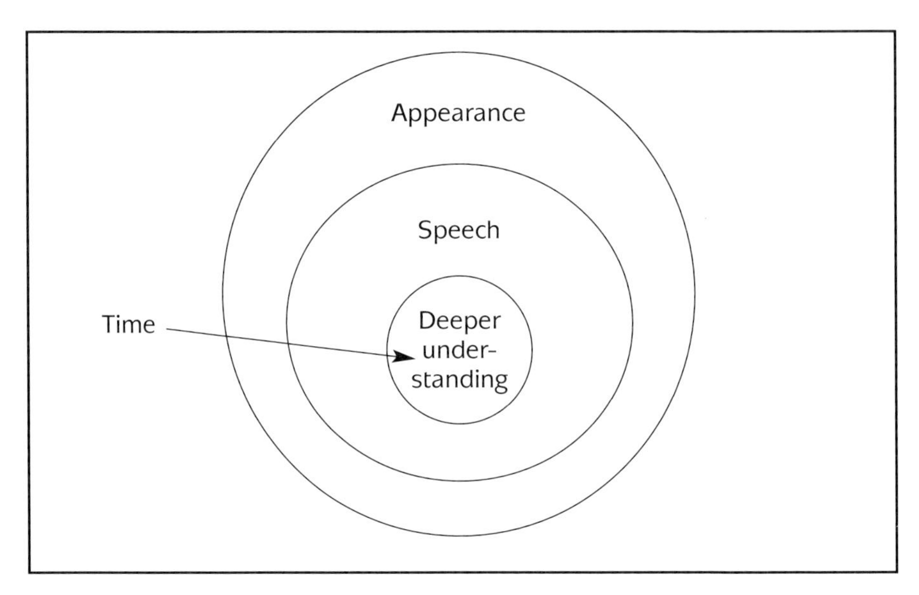

Figure 10 First impressions

THE STRUCTURE OF THE INTERVIEW

A well-structured interview has certain advantages. It will:

- provide order and discipline;
- help the panel cover the agenda – nothing will be missed;
- assist the panel with time-keeping; and
- provide a focus to proceedings.

You will probably find that the interview will start with a brief welcome and settling-in period. The chairman will then ask you to give your presentation. This will be followed by approximately 10 minutes of questions on the issues raised. Following this, the panel will move on to consider your wider areas of training and experience in relation to your chosen competencies, with the chairman and assessors taking it in turn to question you. This part of the interview may last for up to 35 minutes. The chairman will then spend a few minutes closing the interview and will give you the opportunity to have the last word.

Note that the structure of the interview is somewhat different for candidates taking the Antiques and Fine Arts route, as explained above. Candidates for this route are strongly advised to obtain further details of their additional requirements from the RICS Practice Qualifications Department.

THE PRESENTATION

Your presentation will be based on your critical analysis. The objective of the presentation is to give the panel an outline of the purpose, investigations and conclusions concerning the work detailed in this report. It should last for 10 minutes.

I should stress that you are not expected simply to read from a prepared speech. Rather, the panel will judge you on your presentation skills, as well as on the content of the presentation itself.

Following the chairman's introduction, you will be asked to give your presentation. You will be allowed either to sit or stand. The panel will not, in normal circumstances, interrupt you during the 10 minutes, although the chairman may let you know that you are approaching the end of your 10 minutes – indicating that it is time to draw your presentation to a close.

You should note that overhead projectors and screens will not be available at the assessment centre. However, you can use a stand-alone laptop, if the format is appropriate for an audience of three. My personal view is that laptops are not appropriate. The presentation is more of a 'sit opposite and talk you through' than a 'stand and present to a large audience using IT' type of situation. Before you call me old-fashioned, think about whether a lot of technology is really appropriate for a short presentation to three people, in an 'office' type situation?

After your presentation a 10-minute slot is allocated for questions on this. The chairman may simply set the scene and then split the time 50:50 between the two panel members, or he or she may also wish to ask some questions, and will allocate time to do this.

When the interview has been drawn to a close, the panel will discuss your overall performance. They will consider whether your presentation should pass or be referred.

Criteria applied by the panel

A further series of criteria have been set to assess this part of the interview. During your presentation, you will be expected to demonstrate:

- good oral communication;
- presentation skills, i.e., eye contact, the appropriate use of body language, and voice projection;
- clarity of thought (that is, your presentation should have a good structure).

The above summarizes how the panel will operate and what they are looking for. I now want to give you some practical guidance, in 10 simple steps, concerning the presentation. Consider the following aspects:

1. stage presence: all presentations involve a certain amount of acting/theatrics. Try to imagine that you are on a stage, giving a performance;
2. preparation and rehearsal: prepare, plan and rehearse. Write your presentation out in full so that it can be read in around 13 to 14 minutes (you will not be reading verbatim on the day and will therefore naturally cut it down to around 10 minutes). Rehearse in front of colleagues, family and friends. Ensure that the first time you give your presentation is not at your final assessment.

You've got plenty of time to prepare, so make sure that you can deliver it backwards, standing on your head, in 9 minutes and 59 seconds!

3 structure: think about the structure of your presentation and break it down into manageable chunks. Consider the following five 'Ps':
   - position: introduction – your name and outline of project, etc.;
   - problem: key issues specific to the task;
   - possibilities: lateral thought, options, why some options were rejected, etc.;
   - proposal: option adopted, critical appraisal, lessons learnt and closing remarks;
   - preparation: forward plan before the day, and rehearse, rehearse, rehearse!

   You will note that the above fits neatly in with the structure of the critical analysis. By following this structure you should achieve the 'clarity of thought' criterion that the panel is looking for;
4 make the presentation interesting: you only have 10 minutes, so do not laboriously repeat the details of a property, its construction, and so on. Get down to the problems and your solutions, i.e., the things that will interest the panel;
5 the audience: it is important that you consider your audience. Think about how you will manage proceedings, the possible use of a short hand-out, and so on;
6 key sentences: use sign-posting at the beginning and end of each section of the presentation. For example, 'I would like to start by giving the panel a brief overview of my presentation' and, 'That concludes my opening remarks and I am now going to move on to the key issues and problems that I was faced with ... ';
7 pauses: don't race through the presentation at high speed. Think about your pace and use pauses to denote the natural breaks in the structure and to signify the next chunk or section of your text;

8 visual aids: have a series of bullet points to jog your memory on a single sheet of paper or, if you need a little more, write the main headings with a series of bullet points underneath on postcards, which you can flick through during the presentation. You might also find a desktop flip chart useful;
9 body language, voice and eye contact: whether you sit or stand may be dictated by the size of the room and the distance between you and the panel. It is very off-putting if you choose to stand and are very close to the panel – you will give the impression of towering over them. If you are unable to stand well back, take the easy option and sit.
–think of how you use your voice in terms of volume, pace and tone to emphasize issues or changes in direction.
–eye contact is also important. If you are working from notes, make sure that you keep looking up and making eye contact with the chairman. Every once in a while let your eye contact 'sweep' the other assessors;
10 your appearance: don't forget to dress for success. Make sure that you feel confident as you walk through the door into the interview.

QUESTIONING TECHNIQUE

Questioning technique is a key issue in any interview. It is through questioning that the interviewer will gather the information needed to make a decision, based on the criteria, as to whether to employ, promote, dismiss, or in our case, pass the APC. It is important that you understand how the interviewer will operate and what techniques will be employed. Therefore in this section I have provided an insight into the various skills that will be exercised by an interviewer.

During the course of the interview the assessors will mainly use open questions. An open question is where a sentence starts with the words: what, why, when, how, where or who. This allows you to expand upon your answers and provide the information that is required to assess your competence.

There may be occasions when the interviewer will use closed questions. A closed question will elicit a yes/no or one-word type of answer and will be used to clarify facts or information.

Assessors will ensure that questions are well-phrased, unambiguous and concise. They will usually ask one question at a time and use short scenarios and narratives to help you understand the question. Supplementary questions will be used to probe and test your problem-solving ability and depth of knowledge. They may also be used to help and encourage you in moments of difficulty.

So far as practical guidance on questioning technique is concerned, you may find figure 11 useful.

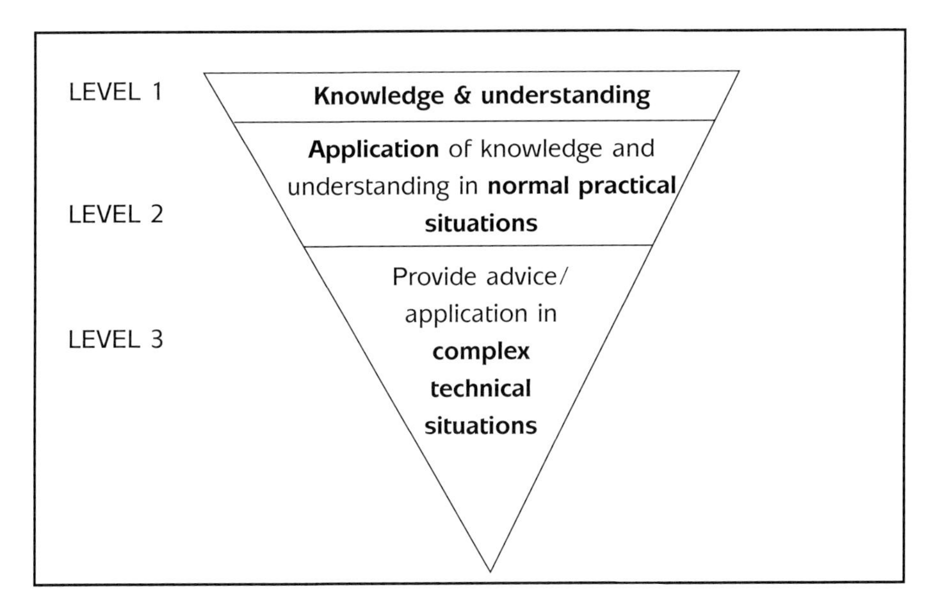

Figure 11 Questioning technique

The concept behind the inverted triangle is to signify the probing of the depth of your knowledge, skills and ability. Think in terms of three progressive levels of questioning:

- level 1: tests your knowledge and understanding of the theory behind any of the competencies;
- level 2: tests your knowledge and understanding in the context of the application of this theory to normal practical situations. Assessors will do this by targeting your training and experience;
- level 3: concerns your experience and ability to provide reasoned advice and reports to clients. There may also be reference to more advanced technical issues. This will be done with regard to your experience, but there may be elements of questioning at this level that are outside your direct training and experience. This is an example of where the panel may 'step outside' into the part competency areas. This is covered in more detail below.

As an example, the following is a typical line of questioning from the Rural Practice route. It is based on the agriculture competency (Ref 003):

- 'I notice from your experience that you have carried out a rental valuation of Quarry Hill Farm. How did you go about the inspection, particularly in relation to the farm buildings, land and crops?' (level 1);
- 'In preparing your valuation, and over and above these physical factors, what part did budgets and profitability play? How did these matters affect your thinking and approach to the valuation?' (level 2);
- 'I note that your client in the situation was trading at a loss. What factors did you consider or look at before advising on how to proceed with changing the business in terms of crops grown, overheads, marketing and general management?' (level 3).

An alternative approach to level 3 might test the provision of advice to the client:

- 'What issues and headings did you cover in your valuation report to the client?'

It is important to remember that in some of the optional competencies you are able to select levels. If competence is only required to level 1 or 2, the assessors' questioning will not probe beyond that point. However, don't forget that the final assessment is a part competency-based interview. At certain points, you may be tested outside of your main areas of training and experience. Therefore, another approach at level 3 in the above example could be for the panel to step outside of the candidate's experience. In this scenario, the questioning concerning trading at a loss could start, 'Assume that your client was trading at a loss ... what advice would you give?'

I would now like to return to and expand your knowledge and understanding of the concept of the part competency-based interview. The reasons for this approach are explained in chapter 4. In this chapter, it might be useful to give you some further guidance, particularly concerning the differences in approach to questioning that you will encounter.

The phrasing and approach to questions that are purely targeting your competency may be as follows:

- 'Give me an example of...'
- 'I see from your final record of progress that you have...'
- 'What was your role/involvement in...'
- 'Outline the process/procedures that you employed/adopted...'
- 'What problems did you encounter...and how did you solve them...'

- 'What was the outcome...'
- 'What did you learn...'

Problem solving and learning is a key feature of competency-based interviewing, so be prepared for this line of questioning.

The questions relating to the 'stepping outside' of your experience, i.e. the part competency approach, may be phrased as follows:

- 'What would you do differently next time...'
- 'How would you apply this experience in tackling...'
- 'What if a situation/problem arose...'
- 'You have not had specific experience of...so in theory how would you...'
- 'Where might you seek further advice and guidance...'

This part competency-based approach is very much geared to the 'big picture' so far as the objective of the APC is concerned – assessing that you are 'competent to practise'. It looks to test skills transfer and to assess your ability to take learning and experience in one area of practice and apply it to another. In addition, it seeks to establish that you are aware of your limitations – hence the line of questioning concerning seeking further advice and guidance.

The above outlines the approach to questions relating to the core and optional competencies. You should also be aware of the approach that will be adopted to test the mandatory competencies.

Ordinarily, when testing mandatory competencies, the assessors will refer to your past experience, rather than adopting a theoretical, textbook approach. Because of time constraints in the final assessment interview, assessors have been advised to keep the time spent on these competencies in proportion. A large part of the interview will be devoted to the technical aspects of the core and optional competencies. You will find that very often the demonstration of your ability in the mandatory competencies will be woven into the main fabric of the interview, and will be drawn out of your responses to the technical questioning in your core and optional competencies, as well as from your written reports and presentation. However, there will be occasions when you will face some specific questioning. A typical approach is for the panel to target the professional development that you have carried out in order to develop knowledge and understanding at level 1 in any of the mandatory competencies.

Interviewee technique

I will now give you some practical guidance on how to be a good interviewee. Once again, I have devised a simple 10-point guide:

1 ice-breaker: make sure that you prepare for the ice-breaker in advance. Give it some thought and use it to provide the panel with some further insight into your training and experience;
2 pause for thought: before responding to questions, always pause for thought. The panel will not be expecting you to leap into answers. Always stand back for a moment, consider the question, collect your thoughts and then deliver your response – look before you leap;
3 active listening: think of listening in an interview as an active, rather than a passive, skill, and make sure you concentrate on what is being said. In an examination, you would read a question and then read it again, to stimulate and recall information relevant to the subject. In an interview, you will usually have only one chance to hear the question, so practice repeating the words in your mind as it is being asked by the assessor. This will help you commit the question to memory and assist your powers of recall when answering it;
4 repeat the question: if you are not entirely sure what the question was, ask the assessor to repeat it. This is a useful technique to employ if you are asked a long-winded and

complicated question – it will force the assessor to revisit the phrasing of the question and may help to focus your thoughts. However, you should do this only sparingly;

5 chronological order: think about chronology when giving your response. A lot of the questions you will be asked will command a response with a natural chronological order, for example, relating to techniques employed or the order of carrying out a particular task or function. This will help you collect your thoughts and ensure that you do not miss any important aspects of the response required;

6 key issues: if there is no chronological order inherent in your response, you may prefer to think of your answer in terms of key issues or bullet points. In this respect, you may find it helpful to use the various competency statements as a guide to your preparation for the interview. These statements will provide the trigger for the areas of questioning by the panel;

7 unfamiliar areas of experience: in the interview you will be asked questions outside of your main areas of experience. In these instances, do not be afraid to qualify your response. Don't forget, the assessor may have placed you outside of your main areas of experience in order to test how well you can draw from these areas and apply your learning and knowledge to an unfamiliar territory. Make sure you qualify your answers by indicating whether you have only limited or no experience in any particular subject area;

8 mental blocks: everyone suffers from nerves during an interview situation. There may be occasions when you have a complete mental block or stumble with the answer. Don't worry, the panel are trained to help you in these situations, for instance, by giving you more time, or approaching a question from a different angle. You will also be offered the opportunity to return to areas of questioning at the end of the interview;

9 bluffing: do not try to bluff or waffle your way through any of the answers. The assessors will all be experienced surveyors and will probably detect when you are unsure or are attempting to

guess the answers. Often, when we are unsure in an interview, our body language or tone of voice can be a give-away. Bluffing will also give a bad impression in terms of duty of care and being aware of your limitations. If you are not sure of an issue, or really have no idea of the response required, simply tell the panel;

10 final word: you will be offered the last word by the chairman when the interview is drawing to a close. At this stage only re-open an area of questioning if it is absolutely vital. Close the interview by thanking the assessors for their time, smile and leave the room.

NOTE TAKING

The assessors will take brief notes during the interview, to ensure that they are being fair to candidates. Do not be unnerved by this. It is being done for your benefit, to ensure that the interview is a fair and impartial assessment of your performance.

The notes taken will, in the final analysis, assist assessors in considering whether to pass or refer you. In the event of a referral, the notes will be an invaluable source of information and guidance for you.

EQUAL OPPORTUNITIES

The need to be conscious of equal opportunities crops up in many aspects of the final assessment process. The following is a summary of the main points:

- the assessors will keep a record of the interview;
- the chairman will handle the opening five minutes with great care to help settle your nerves;
- the chairman will ensure that you are 'fit and well and ready to proceed'. He or she will check that you are not suffering from ill health;
- the chairman should control time across the interview as a whole and within the various components;

- the assessors will link questions to your training and experience in terms of the mandatory, core and optional competencies;
- questioning will mainly relate to the criteria set down for the APC;
- the chairman will give you the last word before drawing the interview to a close.

CONDUCT, BEST PRACTICE AND CUSTOMER CARE

The panel will think of you as its 'customer' and consider 'best practice' to be about the delivery of excellent customer care. The preparation beforehand, the tone of voice used when asking questions and the way that the assessors look at you during the interview will all affect how you feel and therefore perform.

In an interview situation, these matters are important to both the interviewer and interviewee. Let us now consider some of these issues from both sides of the interview table:

- room layout: the chairman will ensure that the environment within which the interview takes place is prepared to your advantage. He or she will ensure, for example, that you will not be staring into the sun. The chairman will check that the panel have name cards in front of them so that you are not forced to try to remember names from the introductions. Before you enter the room, think about how you will greet the panel. As soon as you enter the room, decide where you will put your plans, papers, etc.;
- eye contact: you must always remain attentive and appear interested. However, do not spend the whole time 'eyeballing' the chairman or one of the assessors. Try to think about looking just over the interviewer's left shoulder, right shoulder or just below the chin. This will take the sting out of the eye contact, but will still signal to the panel that you are focused and attentive;
- 60 minutes is a long time to remain attentive. It is easy to gaze out of the window for a few minutes, read from some of the

details in front of you, or fidget. Stay conscious of the impact that such actions may have on the panel;

- body language: you must be aware of the impact that body language can have on your performance. Be conscious of your facial expressions – a pained look may immediately convey the impression that you are finding an area of questioning difficult. Never underestimate the importance of a smile in terms of building a rapport with the panel. Also think about how you sit. Avoid sitting with your arms folded, or leaning forward, eyeball to eyeball, and speaking in a harsh or agitated tone of voice. This will not give the panel a good opinion of you;
- voice projection: your tone needs to be warm and enthusing, the volume positive and confident and the pace such that the panel is able to follow and understand your answers;
- listening skills: your ability to listen and concentrate during the course of an interview is important for two reasons. First, to ensure that you interpret the interviewer's questions correctly; and second, to help you convey interest and attention through positive body language;
- closing the interview: before drawing the interview to a close, the chairman will ask if you have anything else to say. This opportunity will be kept within the parameters of the interview. The chairman might say, 'Is there anything we have discussed that you would like to add to or clarify?', thereby avoiding re-opening the interview on an entirely new topic. After this, the chairman will probably say, 'The interview is now finished. Thank you for attending'. He or she will stand, perhaps give a final shake of the hands, and gesture or lead you to the door. It is important that you leave on a positive note, so try to smile and don't forget to say thank you.

SUMMARY

- An interview is an information-gathering process, with the aim being to provide the interviewer with the information that will satisfy the objective and criteria for the interview.
- Rehearse your opening lines so that you get off to a good start. This will help with your nerves.
- Dress for success. If you look good, you will also feel good.
- If you are unwell, advise the RICS staff at the assessment centre, who will provide you with any assistance required.
- Practice your presentation extensively before the day, so that you can deliver it with confidence.
- In discussion with your supervisor, counsellor or colleagues, try to anticipate the questions that may be asked on your presentation.
- Focus on the criteria that have been set for your presentation: good oral communication skills, presentation skills and clarity of thought.
- Understand the difference between open and closed questions and practise responses that make use of chronological order or summarize key points.
- Practise active listening.
- Use the competency statements while preparing and revising for the interview. These statements will act as the focal point for the questions from the panel.
- Do not be put off by the panel taking notes. This is for your benefit, to ensure that the final assessment is fair and is based upon the evidence presented by you.
- At the end of the interview always smile and thank the panel for their time.

Appraisal, referral and the appeal system

This chapter will consider what happens after the final assessment interview. It will look at how the assessment panel reach their decision and will also consider referral reports and how you can make an appeal.

CANDIDATE APPRAISAL

Before the final assessment interview commences, the chairman will agree with the panel that there should be four or five minutes of silence after you leave the room. This period of quiet reflection allows panel members to review notes, events, answers and to start evaluating performance generally, so that a balanced appraisal may be made. A large part of this process will involve the panel members considering how the details of your training and experience have stood up to questioning during the interview. As a starting point, they will consider your performance against the five broad headings that equate to the concept of being 'competent to practise' (see chapter 1). The assessors will link the mandatory, core and optional competencies to these five headings to arrive at their decision. There will of course be a great deal of overlap – for example, you will demonstrate your oral communication skills each time you answer a question. Figure 12 indicates the mental links the assessors will make between the competencies.

It is therefore vital that you have a clear picture in your mind of what being 'competent to practise' looks like from the outset. Your goal should be to demonstrate these five areas of skill and ability during your final assessment interview.

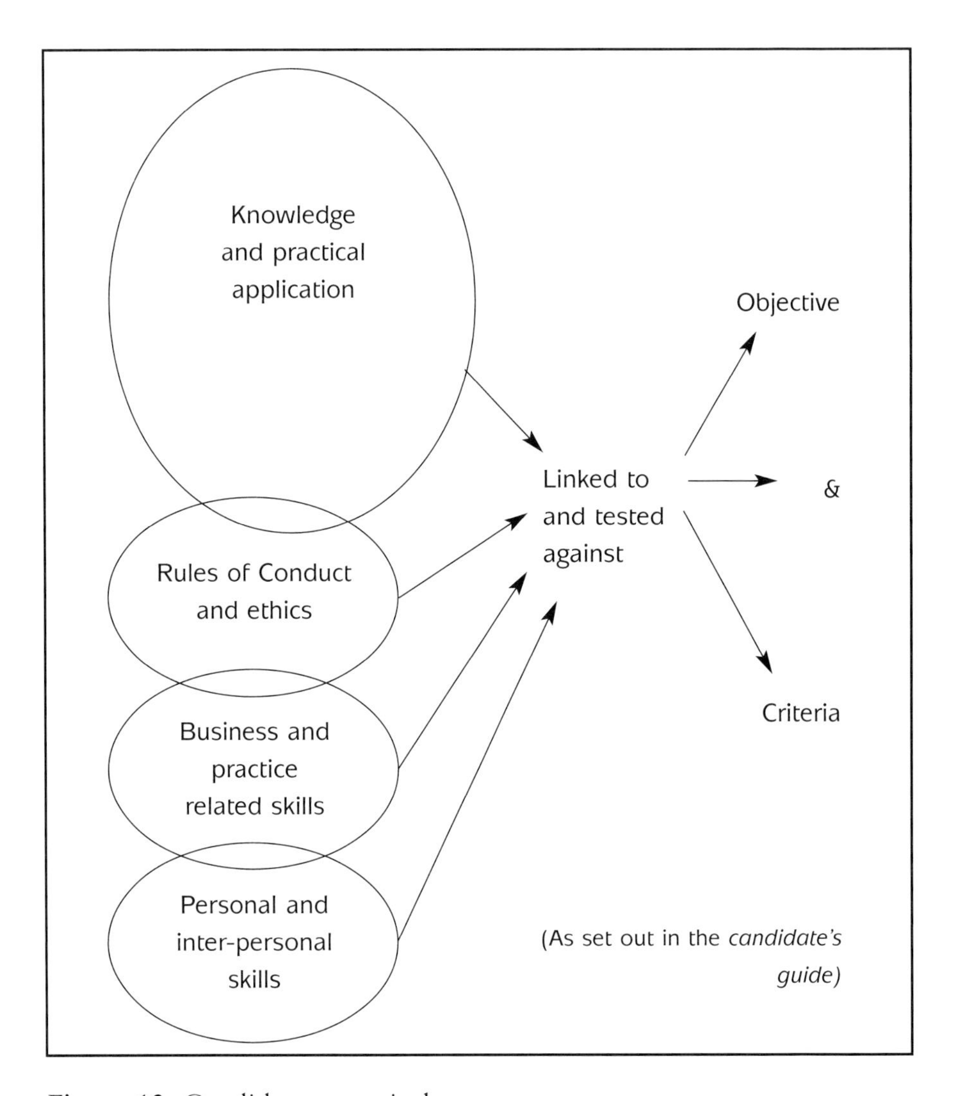

Figure 12 Candidate appraisal

As well as questioning you on your training and experience, the panel will check to ensure that you meet the minimum training requirement of 400 days of experience gained within 24 calendar months.

The panel will then move on and consider your critical analysis and presentation. These will be considered against the criteria set for these components of the final assessment (see chapters 4 and 5). Finally, they will consider your interim and final assessment records. They will also check that your professional development meets with the minimum requirement of 48 hours per annum.

DRAWING THE SIX COMPONENTS TOGETHER – THE HOLISTIC VIEW

The panel will draw the six components of the final assessment together on the front page of the marking form, as shown in figure 13.

There are three columns on the form: refer, marginal and pass. This allows the panel to take a holistic view of your performance and then make a decision as to whether you are considered 'competent to practise'. All candidates are different and skills will vary. Some candidates will show strength in technical and professional matters and perhaps be a little weaker with inter-personal and communication skills. With other candidates, the opposite may apply. The panel will take an overall view of your performance. However, if any of the six components fall so far short of the criteria that the refer box is ticked, you will be referred overall and will have to retake the whole assessment.

THE ROYAL INSTITUTION OF CHARTERED SURVEYORS

Affix Photo Here

ASSESSMENT OF PROFESSIONAL COMPETENCE

SPRING/AUTUMN ....

Name of Candidate ....

Number .... Age ....

Firm's Name ....

Previously Referred YES/NO

Specialist Area ....

I commenced the APC before January 1997 YES/NO (delete as applicable)

FOR OFFICE USE ONLY - PLEASE DO NOT WRITE BELOW THIS LINE

Currently recording ☐

Experience route ☐ (1 year monitored experience)

Diary exempt ☐ (No diary required)

Mature Entry ☐ (1 year monitored experience)

Special consideration ____________________

FINAL ASSESSMENT

Overall Assessment	Refer	Marginal	Pass
Training and Experience			
Interim and Final Assessment Records			
Professional Development			
Critical Analysis			
Presentation			
Interview			

Overall Result: REFER/PASS (delete as necessary)

Chairman (Block Capitals) ....

Signature .... Date ....

Assessor .... Assessor ....

Figure 13 Marking form

Role of the chairman

During this decision-making period, the chairman will ensure that the completed forms represent a consensus of opinion and the panel's final view. He or she will listen to the panel members, particularly with regard to their areas of specialism, and will weigh and balance their views to ensure that the final decision is fair.

THE FINAL DECISION

It should be remembered that you will not be expected to demonstrate a level of knowledge and experience equivalent to that of an experienced practitioner. The panel's decision will be based on you having demonstrated competence to the required level, tested against the various criteria that have been set. The panel will not be expecting a level of knowledge equal to that of their advanced years!

It is also worth noting that there are no quotas or pass rates for the APC. The benchmarks that the assessment panel will work to are the levels of competence indicated in the *APC/ATC requirements and competencies* guide, tested against the various criteria set out in the *candidate's guide*. All candidates that meet these requirements will pass the final assessment.

Results will be sent to you by first-class post within 21 days from the date of your interview. Remember to let RICS know if your correspondence address changes. A pass list will also be published on the RICS website after each assessment period. For security reasons, no results will be given over the telephone, by fax or to a third party.

REFERRAL REPORTS

In the event of a referral, the chairman of your assessment panel will provide you with as much information as possible. The referral report will be linked to the six components of the final assessment: the training period; interim and final assessment records; professional development; critical analysis; presentation; and interview. The panel will have been asked to give reasons as to why you did not meet the required levels of competence. These should be given with reference to the specific competency and with a clear explanation of how you have fallen short of the criteria. You may also receive advice on further training and development that has been identified as suitable by the panel. Finally, you may receive feedback on aspects of your performance that were satisfactory. The objective of this is to enable you to address any weaknesses before the next assessment and to be successful when you re-sit.

The receipt of a referral report will be a miserable experience. However, you must remain positive and focused. Go back to some of the key concepts behind the critical analysis, be introspective, learn from your mistakes and improve areas of weakness. You have had the benefit of the experience, so learn from it and build upon your strengths, to enable you to be successful on the next occasion. Be philosophical. It may be that you were not quite ready for the final assessment.

If you are referred, there are some minimum requirements that you must satisfy before you re-sit the final assessment:

- record a minimum of a further 100 days of relevant professional experience. The assessors will probably give you guidance on this;
- undertake a minimum of a further 24 hours of professional development;
- write a new critical analysis; or if recommended, resubmit the original, suitably amended and updated;
- complete a summary forward plan to cover the further 100 days of training and experience. After this, you must complete a summary of progress form. Another supervisor's and counsellor's report form must also be completed. All of these forms can be found at the back of the *candidate's guide*;

- submit a copy of your record of progress giving details of the further 100 days of training and experience in relation to the competencies;
- be re-interviewed and give a presentation on the relevant critical analysis.

APPEALS

If you are unsuccessful and are in any way aggrieved by any aspect of the final assessment, you can make an appeal. This must be received by RICS no later than 14 days from your result being posted to you. The details of how and when to lodge an appeal will be sent with the referral report and application for reassessment.

If you do feel aggrieved, my advice is as follows:

- it is very difficult to get a feel for the outcome of an interview immediately afterwards. Your nerves and adrenaline level will still be high and you will find it difficult to be truly objective;
- if you are still feeling aggrieved when you get home, sit down and write a list of the aspects of the interview that went well, and then make a list of the aspects that did not go well, noting why you feel unhappy;
- after 24 hours, go back to your notes and consider whether there are any aspects you still consider to be unfair.

When you receive the result, you may be pleasantly surprised. I remember receiving telephone calls from two candidates immediately after their assessments claiming that they had been unfair and wanting guidance on how to appeal. Shortly afterwards they received their results and both rang me to say, rather sheepishly, that they had passed. So do not over-react to one or two aspects of the assessment that did not go so well. The tendency in a highly charged interview situation is to remember only the downsides.

In my view, there are three main reasons why an appeal might be validly made:

1 administrative or procedural matters: the panel may not have been provided with the correct information and detail; or something may have gone wrong, for example, denying you the opportunity to make your presentation;
2 the questioning and testing of competence concentrated too heavily outside of your main areas of training and experience;
3 any form of discrimination.

Hopefully, such problems will not arise. In case they do, however, RICS has established an appeal system.

THE APPEAL SYSTEM

If you do want to make an appeal, it is important that it is made by you, and not a third party. You must clearly state the grounds upon which your appeal is made and provide any necessary supporting evidence.

After some additional research has been carried out by RICS, your appeal will be considered by the chairman of the Practice Qualifications Group and two nominated members from the faculty of your chosen route.

If your appeal is successful, your fee will be returned and a reassessment, using all of the original paperwork, will be arranged as soon as possible. You will be reassessed by a different panel and the normal rules and procedures will apply. Conversely, if your appeal is rejected, you will be eligible for reassessment along the lines of the notification that was sent with your referral report.

SEEKING ADVICE AND FURTHER GUIDANCE

- Be introspective and do some soul-searching. Is there anything you can learn from a referral to make you successful on the next occasion?
- With your referral report at hand, talk to your employer, your supervisor and your counsellor.
- You may also wish to approach your RICS training adviser (RTA), APC doctor or RICS itself.
- Consider how additional professional development could assist you in addressing any of the shortcomings noted in your referral report.

SUMMARY

- Ensure that your paperwork and final submissions are in order. The panel will check to ensure that you meet the training and experience requirements (a minimum of 400 days within 24 calendar months), as well as the final assessment record and professional development requirements.
- In drawing the six components of the final assessment together (training and experience; interim and final assessment records; professional development; critical analysis; presentation; and interview), the panel will take a holistic view of your performance.
- The decision will be arrived at by considering the views and opinions of all three panel members. This is to ensure a fair and balanced outcome.
- The six components of the final assessment will be weighted, but all the assessment criteria must be met.
- The panel will not be working to quotas or pass rates.
- If you are unhappy with the interview and wish to make an appeal, try to be objective. Seek advice from your supervisor and counsellor.
- If you wish to make an appeal, you must do so no later than 14 days from your result being posted to you.

Conclusion

I would like to conclude this book by reiterating some of the comments I made in chapter 1. The APC is first and foremost a period of training and practical experience. If you follow the guidance and ensure that you are learning the skills and reaching the levels of attainment required in the various competencies, you will be well on your way to a successful final assessment interview. If the training and experience has been correctly put in place over the two-year period leading up to the final assessment, the outcome should be a formality.

It is when candidates have not followed the various guidance that is available, and have not remained focused on the competency requirements, that the final assessment results in an unsatisfactory outcome.

It is also essential that you focus on the various skills required for the final assessment: report writing, presentation and interview skills. The practical guidance set out in the previous chapters should provide you with some insight to the challenges that lie ahead, not just with the APC, but for the rest of your career.

Finally, the following is a list of vital points to be borne in mind as you progress through the APC:

- you cannot start recording experience until RICS has accepted your application for enrolment;
- if you change employer during the training period, you must inform RICS;
- the training period is a minimum of 400 days within 24 calendar months;
- professional development must comprise a minimum of 48 hours per annum;
- the interim assessment must be completed within one month of the first 12 months of training. You must then complete a further 12 months of training and experience before you are eligible for the final assessment;
- you must have signed off the appropriate number of competencies, and to the levels required by your route, before you are eligible for the final assessment;
- the final assessment is a part competency-based interview, so be prepared for questions outside of your main areas of training and experience. Remember, you are being interviewed to test whether you are 'competent to practise' overall as a chartered surveyor;
- don't forget your supervisor and counsellor declarations on the:
  - training agreement;
  - competency achievement planner;
  - diary;
  - log book;
  - record of progress;
  - interim and final assessment records; and
  - critical analysis.
- stay focused on the competency statements for your route, particularly the criteria that overarch the assessment, as well as those that apply specifically to the critical analysis and presentation;
- finally, you have 14 days from the date on which your results are posted by RICS in which to make an appeal, should this be necessary.

I would like to offer you one final thought based upon my 27 years' experience as a chartered surveyor – and that is that your APC never ends. Your professional competence will continue to be assessed by employers, clients and peer groups throughout your career. As the world around you changes in terms of consumer demands, law and technology, the content and focus of your training and development will also change.

The most recent development in this respect is the introduction of the Certificate of Management Studies (CMS). From September 2004 all candidates who pass their APC must complete their CMS within five years (further details are available from RICS).

Therefore, when you become chartered, think of the RICS programme of continuing professional development (CPD) and the CMS as vehicles to assist you with your essential and ongoing training and development needs, in a rapidly changing environment.

Index

Index